COMPREHENSIVE ANALYTICAL CHEMISTRY

ELSEVIER SCIENTIFIC PUBLISHING COMPANY
335 JAN VAN GALENSTRAAT
P.O. BOX 211, 1000 AE AMSTERDAM, THE NETHERLANDS

Distributors for the United States and Canada:

ELSEVIER/NORTH-HOLLAND INC.
52, VANDERBILT AVENUE
NEW YORK, N.Y. 10017

LIBRARY OF CONGRESS CARD NUMBER: 58-10158

ISBN 0-444-41732-X (Vol. IX)
ISBN 0-444-41735-4 (Series)

WITH 93 ILLUSTRATIONS AND 16 TABLES

PRINTED IN THE NETHERLANDS

COMPREHENSIVE ANALYTICAL CHEMISTRY

ADVISORY BOARD

Contributors to Volume IX

R. Browning, Department of Pure and Applied Physics, The Queen's University, Belfast, N. Ireland

S. Hofmann, Institut fur Werkstoffwissenschaften, Max-Planck-Institut, 7000 Stuttgart 1, B.R.D.

P. Tschöpel, Laboratorium für Reinstoffe, Max-Planck-Institut, 7070 Schwäbisch Gmünd, B.R.D.

Wilson and Wilson's

COMPREHENSIVE ANALYTICAL CHEMISTRY

Edited by

G. SVEHLA, PH.D., D.SC., F.R.I.C.

Reader in Analytical Chemistry
The Queen's University of Belfast

VOLUME IX

Ultraviolet Photoelectron and Photoion Spectroscopy
Auger Electron Spectroscopy
Plasma Excitation in Spectrochemical Analysis

ELSEVIER SCIENTIFIC PUBLISHING COMPANY
AMSTERDAM   OXFORD   NEW YORK
1979

WILSON AND WILSON'S

COMPREHENSIVE ANALYTICAL CHEMISTRY

VOLUMES IN THE SERIES

Preface

In *Comprehensive Analytical Chemistry*, the aim is to provide a work which, in many instances, should be a self-sufficient reference work; but where this is not possible, it should at least be a starting point for any analytical investigation.

It is hoped to include the widest selection of analytical topics that is possible within the compass of the work, and to give material in sufficient detail to allow it to be used directly, not only by professional analytical chemists, but also by those workers whose use of analytical methods is incidental to their work rather than continual. Where it is not possible to give details of methods, full reference to the pertinent original literature is made.

Volume IX contains three chapters, all connected with fast growing branches of spectroscopy. The first, on Ultraviolet Photoelectron and Photoion Spectroscopy, presents the theoretical and practical foundations of the technique. We intend to publish follow-up chapters on its analytical applications (ESCA) and on X-ray photoelectron spectroscopy in a later volume. The second chapter, entitled Auger Electron Spectroscopy, is closely related to the first, describing a powerful method for surface analysis. In a future volume, a contribution on electron microprobe analysis is planned, thus making the field of surface analysis more complete. The last chapter in this volume on Plasma Excitation in Spectrochemical Analysis deals with a topic which revolutionized the field of analytical emission spectroscopy. When reading this chapter, the reader will find it helpful to refer back to Volumes IV and V of the series where spectroscopic instrumentation and emission spectroscopic analysis in general have been described respectively.

Our three contributors come from European academic and research institutions and are internationally known experts in their fields.

Dr. C.L. Graham of the University of Birmingham, England, assisted in the production of the present volume; his contribution is acknowledged with many thanks.

November, 1978   G. Svehla

ix

Contents

Chapter 1

Ultraviolet photoelectron and photoion spectroscopy

R. BROWNING

Symbols and abbreviations

The fundamental constants and electrical and spectroscopic quantities are labelled according to S.I. or I.U.P.A.C. recommendations. Similarly, S.I. units are generally used except for the electron volt which is used as the unit of energy to conform to general practice. A tilde mark is used to indicate a molecular ionic state thus: X, A, $\tilde{a}$, etc. Quantities defined in the text at the point of use are not included in the following list.

e (e)	electron (or electronic charge)
g	polarization factor
g_i	statistical weight
i, f	initial, final state labels
r	internuclear separation
CNDO	complete neglect of differential overlap (theory)
D	dispersion
E	energy (of photoelectron)
EC	electron correlation
$G, G(r, E)$	electronic transition moment
I_{if}	ionization potential; i $\rightarrow$ f
I_A, I_V	adiabatic, vertical ionization potential
I_ν	intensity at frequency ν
M	molecule
N	number, or flux, of photoelectrons
SCF	self consistent field (theory)
V	potential energy
ρ	resolving power $E/\Delta E_{1/2}$

w	slit width
ω', ω''	vibrational frequencies
ψ_i	wavefunction for state i
ν	frequency (of light)
β	asymmetry parameter
ΔE, $\Delta E_{1/2}$	increment of energy; peak full width at half maximum
ϵ	electric vector

1. Introduction

The experimental methods described in this chapter have given chemists and physicists a new understanding of the structure of atoms and molecules, particularly because of the directness of the approach. For example, before the technique became available in the 1960s, studies of ionization potentials involved the reduction of complex photoabsorption spectra into Rydberg series, or the finding of elusive appearance potentials. Such techniques involved so much uncertainty and complexity that many found the prospect daunting, if not actually repulsive. Molecular orbitals often seemed unreal except to the theoretician; how could the experimentalist visualise them and appreciate their significance? But the technique of photo-electron spectroscopy has enabled the electrons in a molecule to be extracted one by one, so that their orbital energies may be discovered. The gentle photons can probe the molecule and release the electrons with a minimum of disruption. In addition, the separation of the electronic from the vibrational and rotational motion in a molecule can be displayed for all to appreciate.

Although the first results to be published came from Leningrad in 1961 [7] the subject of ultraviolet photoelectron spectroscopy was developed in the form we recognise it today by Turner and his co-workers at Imperial College London [8]. They recognised that a discharge through helium provides the experimentalist with a source of virtually monochromatic radiation of sufficient energy to release electrons from many valence orbitals, and with sufficient flux for the photoelectric effect to be observed and exploited in a manner which would no doubt have surprised Einstein, whose equation forms the basis on which the results are analysed.

It is hoped that this chapter gives some impression of how rapidly and extensively the technique has been developed over some fifteen

2

years. Investigations are now made of the electronic and vibrational energies to be associated with ionic states, and their relation to the molecule from which they are derived; of branching ratios into different ionic states; of the angular dependence of the photoelectrons produced; of ionic dissociation and the theory of mass spectra; and of autoionization and other configuration mixing effects. All these topics are mentioned here, at least to some extent, but the subject is so broad that many aspects are treated superficially. The closely related subject of photodetachment is not treated at all, since electron detachment from negative ions is generally achieved using visible photons. The boundary between ultraviolet photoelectron spectroscopy and the spectroscopy of inner shells is reasonably well defined by the words ultraviolet and X-ray; this chapter will be concerned with photons in the energy range around 10—50 eV.

The general field has been extensively reviewed before, notably in the books and papers listed under general references [1—6]. For this reason, the author has attempted to cover quite a wide range of topics, but with sufficient detail for the fundamentals of the subject to be understood. In keeping with the purpose of this series, the experimental aspects of the subject are rather fully developed. Inevitably, however, much excellent work is not even mentioned, and the illustrative applications reflect the bias of a physicist unaware of many of the most typically "chemical" aspects of the subject: but an attempt has been made to develop the experimental aspects which seem to offer good prospects for future development.

2. Fundamental principles

(A) PHOTOIONIZATION AND PHOTOELECTRON SPECTROSCOPY

When a beam of electromagnetic radiation or photons passes through a substance it will be attenuated by a number of processes. These processes may be characteristic of the individual atoms or molecules, or may also depend on the bulk properties of the sample if the molecules are associated with others as in a liquid or a solid. In this chapter, we are concerned only with the former, *free* atoms and molecules, which are ionized by the photon. The basic process is thus photoionization, which we represent by the equation.

$$h\nu + M_i \rightarrow M_f^+ + e(E) \tag{1}$$

If the ionization energy or potential required to produce the final state, f, is $\epsilon_f - \epsilon_i$, then, since the mass of the ion is large compared with that of the electron, conservation of momentum determines that the recoil energy of the ion is usually quite negligible for photons less than 100 eV. Thus the electron energy E is given by

$$E = h\nu - (\epsilon_f - \epsilon_i) \tag{2}$$

Recognising that the final state of the ion may have vibrational and rotational excitation of a particular electronic state, we can rewrite this as

$$E = h\nu - I_{if} - \Delta E_{vib} - \Delta E_{rot} \tag{3}$$

This is the basic equation used in photoelectron spectroscopy.

The energy, E, probability of ejection (cross-section σ), angle of ejection of the photoelectron, and the excitation or species of the ion, depend on the molecule studied and the photon energy.

The number of photoionization events, N_f, produced in a target of number density n over a distance l is given by

$$N_f = I_\nu \sigma_{if} nl \tag{4}$$

where σ_{if} is the photoionization cross-section to the final state, f, and I_ν is the flux of photons. The total photoionization cross-section

$$\sigma_T(\nu) = \sum_f \sigma_{if}(\nu) \tag{5}$$

may or may not equal the total photoabsorption cross-section, any difference being accounted for by neutral production (e.g. photo-dissociation).

The photoionization cross-section may be calculated theoretically, but the precise formula to be used depends on the normalisation of the wave functions used [9]. For the present purposes, we note that in the electric dipole approximation, the cross-section may be written

$$\sigma_{if}(\nu) = \frac{8\pi^3 \nu}{3c} \frac{1}{g_i} \sum (\Psi_i | er | \Psi_f)^2 \tag{6}$$

where the summation is taken over all the substates involved. The cross-section thus depends mainly on the electric dipole matrix elements. Neglecting autoionization, the cross-section for a particular shell or sub-shell will often decrease from threshold monotonically

4

with frequency (energy). However, $\sigma(\nu)$ may pass through a maximum (Ne $2p$), or a minimum followed by a maximum (e.g. Na $3s$) from a finite value at threshold; but it will eventually decrease with decreasing wavelength, as the photon frequency becomes large compared with the orbital frequency of the electron removed ($h\nu \gg I_{if}$).

(B) THE FRANCK–CONDON PRINCIPLE

Applying this well-known principle to molecular photoionization, we may state that the photoelectron is released in a time which is short compared with a typical vibrational period in the molecule. On an energy level diagram of potential energy V against internuclear separation r, transitions take place vertically upwards. Underlying this principle is the Born-Oppenheimer approximation in which the electronic, vibrational and rotational contributions to the total wavefunction for the molecule (or the molecular ion + outgoing electron) are separated. If rotational states are not resolved as normally occurs in photoelectron spectroscopy, we can see from eqn. (6) that the probability of leaving the ion in the final electronic state e' and vibrational state v' after ionization from the initial state e'', v'' is P where

$$P \propto \langle G_{e'e''} \rangle^2 \left\{ \int \phi_{v'} \phi_{v''} \, dr \right\}^2 \tag{7}$$

If the transition moment $G_{e'e''}$ varies slowly with internuclear separation r and the energy of ejection E of the associated photoelectron, the relative probability for the formation of the various vibrational levels v' of a particular electronic state is determined by the Franck–Condon (F–C) factors as in eqn. (7), i.e. $|\int \phi_{v'} \phi_{v''} dr|^2$ (cf. the intensity distribution in electronic bands, Herzberg [10] Chap. IV, Sect. 4). To illustrate this principle, a potential energy diagram is depicted in Fig. 1 for a hypothetical diatomic molecule AB, initially in its ground vibrational state as in most experiments.

$\widetilde{X} \leftarrow X(v'' = 0)$. The equilibrium internuclear separation for the ground state of the ion is somewhat less than that for the neutral molecule. However, the reduction is not large and consequently the F–C factors are large for the low lying vibrational states of $\widetilde{X}$ only.

$\widetilde{A} \leftarrow X$. The excited ionic state has been formed by the release of a strongly bonding electron. High vibrational states of A are excited: the fourth vibrational level and adjacent levels are prominent. Some A$^+$ ions are expected, since vertical transitions at small r within the

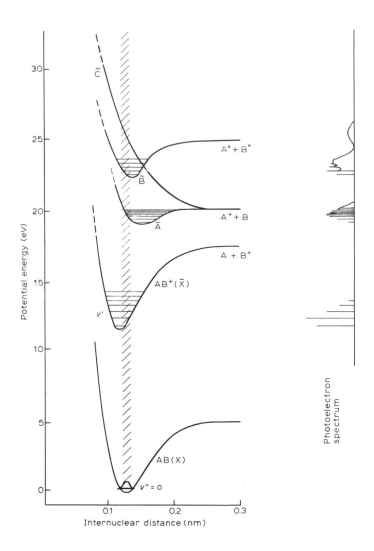

Fig. 1. Potential energy diagram for a hypothetical diatomic molecule AB and some ionic states of AB⁺. The Franck—Condon region is shown shaded, extending vertically from the ground vibrational state, $v'' = 0$, of the neutral molecule; the vibrational probability distribution is indicated for the ground state. The corresponding photoelectron spectrum is shown schematically to the same scale.

F—C region must lead to dissociation. We distinguish two ionization potentials for such a state.

The vertical ionization potential corresponds to the energy required to form the ion in the most probable vibronic level [i.e. the peak intensity of the band in the photoelectron spectrum, but see Sect. 4.C (5)]; I_V.

The adiabatic ionization potential corresponds to transitions to the lowest vibronic level, i.e. $\tilde{A}(v' = 0) \leftarrow X(v'' = 0)$; I_A.

$\tilde{B} \leftarrow X$. The lowest vibrational levels of the $\tilde{B}$ state are excited as before, but the higher levels can predissociate to form A^+ ions. In this illustration, the predissociation occurs rapidly and the vibrational peaks in the photoelectron spectrum are appreciably broadened.

$\tilde{C} \leftarrow X$. A^+ ions are formed through the $\tilde{C}$ state with a wide range of kinetic energies, by direct dissociation. The corresponding photoelectron band will show no vibrational structure and may be difficult to identify as a consequence. In many instances, states such as that shown may lie above more than one dissociation limit, so that photoion spectroscopy must be used to supplement photoelectron spectroscopy if the final ionic states are to be determined.

(C) PHOTOELECTRON SPECTRA: AN ILLUSTRATION

When the energies of the electrons released in photoionization are analysed, a photoelectron spectrum is obtained relating the number of electrons N to their energy E. This photoelectron spectrum will consist of a series of groups of lines (bands), some examples of which

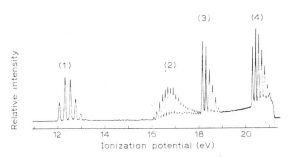

Fig. 2. The 58.4 nm photoelectron spectrum of O_2. For an explanation of bands (1)—(4), see text. (From Samson [5], by courtesy of North-Holland Publishing Company.)

References pp. 82—88

are given in Fig. 2 which shows the 21.22 eV photoelectron spectrum of oxygen.

Following the guide lines established above, we may illustrate the method of interpretation of a photoelectron spectrum using molecular oxygen as an example. We note that the ground state configuration of oxygen can be written

$$KK(\sigma_g 2s)^2(\sigma_u 2s)^2(\sigma_g 2p)^2(\pi_u 2p)^4(\pi_g 2p)^2 \; {}^3\Sigma_g^-$$

and that the vibrational frequency of this state is 1568 cm^{-1} or 194.4 meV. Overwhelmingly the most probable photoionization processes correspond to the simple removal of a single electron from each of the ground state molecular orbitals.

Hence we expect the following states of O_2^+

$$(\pi_g 2p)^{-1} \rightarrow O_2^+(\tilde{X}^2\Pi_g) \qquad \text{Band (1) on Fig. 2}$$

The vibrational spacing in this band $\sim$232 meV which shows, on comparison with the 194.4 meV spacing in the neutral molecule, that the π_g electron is antibonding. The equilibrium internuclear separation of $O_2^+(\tilde{X})$ is less than for $O_2(X)$ and the first five vibration levels are strongly excited in accordance with the arguments given in Sect. 2.B.

$$(\pi_u 2p)^{-1} \rightarrow O_2^+(\tilde{a}^4\Pi_u)$$
$$\rightarrow (\tilde{A}^2\Pi_u) \qquad \text{Band (2)}$$

The vibrational spacing in this band is significantly less than in O_2 and so the $\pi_u 2p$ electron is strongly bonding as expected. Since the two π_g electrons have parallel spins, the unpaired π_u electron in O_2^+ couples to give rise to quartet and doublet ionic states. Of the five final ionic states which are possible from the ionization of a π_u electron in O_2, only the $\tilde{a}$ and $\tilde{A}$ states involve one-electron transitions and are strongly excited as a consequence. The quartet state has the lower energy, in accordance with Hund's rules. It so happens that the electronic and vibrational levels of the $\tilde{a}$, $\tilde{A}$ states overlap so that the two ionic states can only be distinguished in the band by using a high resolution ($\leqslant$15 meV) instrument such as that used by Edqvist et al. [11].

$$(\sigma_g 2p)^{-1} \rightarrow O_2^+(\tilde{b}\ {}^4\Sigma_g^-) \qquad \text{Band (3)}$$
$$\rightarrow (\tilde{B}\ {}^2\Sigma_g^-) \qquad \text{Band (4)}$$

8

These two ionic states are clearly distinguished, and the vibrational spacings in the bands ($\simeq$140 meV) indicate that the bonding properties of the ($\sigma_g 2p$) electron are not as strong as ($\pi_u 2p$). The highest members of the '$\tilde{b}$' vibrational sequence and the whole of band (4) lie above the dissociation limit, and these states do, in fact, dissociate. In this case, dissociation is not evident in the photoelectron spectrum, but photoion mass spectrometry and/or energy spectroscopy may be used to supplement the photoelectron data. If a molecule can perform a sufficient number of vibrations before dissociating, the photoelectron spectra will be unaffected by the subsequent dissociation and, as here, the observed width of the vibrational peaks will not be affected. (If a molecule dissociates rapidly, the vibrational peaks may be broadened, or the vibrational structure removed altogether, as shown in Fig. 1.)

Higher states of O_2^+ corresponding to the removal of electrons of higher ionization potential may be produced, but the above serves as an illustration of some of the most salient points.

(D) PHOTOELECTRON BANDS

In the example given above, fine structure due to spin—orbit interaction in band (2) is not evident because the splitting of the doublet and quartet levels is considerably less than the experimental resolution. However, for atoms and molecules possessing larger internal fields (heavier nuclei) such splitting is commonly observed. Other factors affect the shape or structure of photoelectron bands, and so it seems worthwhile listing the principal factors which determine photoelectron band shapes. These factors (see Fig. 3) are

(a, b) Vibration (and rotation) of molecular ion

As noted before, the rotational structure in photoelectron bands is generally smaller than the instrumental resolution. A notable exception is provided by H_2 where the rotational levels are separated by ~60 meV and rotational transitions may be resolved [Fig. 3(b)].

For most molecules, more than one vibrational mode may be excited so that the vibrational structure is much more complex than that shown here.

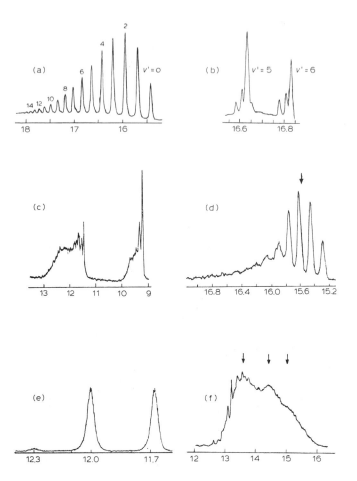

Fig. 3. Photoelectron band shapes (all energies marked in eV). (a) Vibrational structure in H_2: 15 levels can be distinguished. (b) Rotational structure shown by two of the vibrational levels of H_2. (c) The first two bands in benzene, illustrating a gradual loss of vibrational structure due to overlapping of closely spaced peaks. (d) Predissociation in the $\tilde{A}^2\Sigma$ band in HBr. The dissociation limit to $H + Br^+$ $(^3P_2)$ is shown by an arrow. (e) Spin—orbit splitting in the first band of HBr. The $\Pi_{1/2}$ and $\Pi_{3/2}$ components are situated at approximately 12.0 and 11.67 eV. (f) Jahn—Teller structure in the first band in methane. Three overlapping peaks can be discerned, at approximately 13.6, 14.4 and 15.0 eV as indicated by arrows. [(a, b) from Samson [5], by courtesy of North-Holland Publishing Company; (c, d, e) from Turner et al. [6], by courtesy of John Wiley and Sons Ltd; and (f) from Potts and Price [15] by courtesy of the Council for the Royal Society.]

10

(c) Spacing of vibration—rotation levels and band overlapping

If the spacing of these levels is less than the instrumental resolution, which may be quite inevitable, an unresolved band will be obtained with a shape approximated by the Franck—Condon envelope: the discrete levels become smoothed out [see also Sect. 4.C (3)(b)]. Band overlap may also be responsible for such a loss of vibrational structure [12].

(d) Dissociation or predissociation

We have already noted that the dissociation of the O_2^+ $\tilde{B}$ state is not sufficiently rapid for the vibrational energy levels to be perceptibly broadened. However, if dissociation takes place rapidly (in time Δt), the broadening $\Delta E \sim \hbar/\Delta t$ may be appreciable. In the example shown in Fig. 3(d), the vibrational line structure of the $\tilde{A}^2\Sigma$ state of HBr^+ is seen to broaden and eventually to vanish beyond the dissociation limit indicated for H + Br^+ ($^3P_2$) (see refs. 6 and 13).

(e) Spin—orbit interaction

Circumstances such as the removal of an electron from a lone pair will give rise to a molecular ion with an unpaired electron whose spin can couple with the orbital momentum in the molecule to split the final ionic state into two states of different total angular momentum. The effect is generally associated with the presence of a heavy atom in a molecule since the spin—orbit energy is not appreciable for light atoms. In the first band of hydrogen bromide, the spin—orbit splitting is approximately two-thirds of that exhibited by the isolated bromine atom, as may be expected on theoretical grounds [14].

(f) Jahn—Teller and other distortions

Changes in molecular geometry are to be expected on the removal of an electron from a degenerate orbital. Thus in Fig. 3, methane $(1a_1)^2(2a_1)^2(1t_2)^6$ possesses a triply degenerate outer orbital, and the removal of a t_2 electron removes the degeneracy. The CH_4^+ molecular ion that results possesses a complex series of vibrational modes as the molecule, with tetrahedral symmetry initially, adjusts its shape to the loss of the bonding electron [15]. The splitting in such a band caused by these distortions may be substantial, especially for molecules with light atoms and a large force constant. Since it may be expected that the release of an electron from a degenerate orbital will produce an ion in a doublet state, the effect is frequently complicated by spin—orbit coupling, particularly when heavy atoms are involved.

This and other related effects are discussed by Eland [1].

(E) TRANSITION RULES AND INTENSITIES OF PHOTOELECTRON BANDS

As has been mentioned, the most probable photoionization processes involve one-electron transitions, but more complex transitions are not necessarily forbidden. Since the one electron making the transition into the continuum does not carry away a uniquely defined quantity of angular momentum (as an s, p, d ... wave it will carry away $0, 1, 2 \ldots \hbar$ units), the selection rules which play such an important role in photoabsorption between discrete states are not useful when applied to photoelectron spectra. Here we are normally concerned with the initial state of the atom or molecule and the final state of the ion alone (neglecting the free electron), so we can simply indicate two transition rules

$$
\left.\begin{array}{l}
\Delta L = 0, \pm 1 \ldots \pm l \\[2mm]
\Delta S = \pm 1
\end{array}\right\} \text{neutral} \rightarrow \text{ion}
$$

Two-electron processes, in which one electron is ejected and another is excited, are not always "forbidden" in photoionization because the initial and final states may contain elements of the required parity due to configuration mixing. But in ultraviolet photoelectron spectroscopy such processes are usually weak, reflecting a rather small degree of configuration mixing, when compared with the total photoionization cross-section [but see Sect. 4.B (1)]. They should not be discounted, however, in the interpretation of minor structure in photoelectron spectroscopy, or at all if molecular dissociation is the focus of attention.

The intensity of a particular band will often be at or near a maximum at threshold, but eventually it will decrease rapidly with increasing photon energy. Thus, although a comparison between the relative strengths of photoelectron peaks is an aid to orbital identification in similar molecules under similar conditions, such comparisons must be made cautiously. Broadly speaking, the intensities of various photoelectron bands are similar, and while the number of electrons in a particular molecular orbital and the statistical weights of final states may be used as a guide to band intensities, careful attention must be paid to the variations to be expected with the wavelength of the exciting radiation and other effects (see, for example, Price et al. [16]).

(F) ANGULAR DISTRIBUTIONS

The angular distribution of the photoelectrons emitted into a given solid angle with respect to the direction of a beam of photons, which may be polarised, is characterised by an "asymmetry parameter" β. Fig. 4 shows how the intensity of the photoelectron signal varies with angle in two typical experimental arrangements. In both cases, we note that the experimentalist who observes at an angle $\cos^{-1}(1/\sqrt{3}) = 54°44'$ (the "magic" angle) will obtain a signal which is simply proportional to the solid angle viewed and independent of β, a necessary feature if band intensities corresponding to different final states with different values of β are to be compared with one another. Samson [17] has pointed out that the experimentalist may use large solid angles and partially or elliptically polarised light with the same advantage, provided observation is made at an angle for which $\theta = \alpha = 54°44'$. The general expression corresponding to a viewing direction (θ, α) is

$$\frac{N(\theta, \alpha)}{N} = 1 + \tfrac{1}{2}\beta\left\{ \frac{3}{g+1}[(g-1)\cos^2\theta + 1 - \cos^2\alpha] - 1 \right\}$$

where $g = I_x/I_y$. I_x is the intensity of the light with an ϵ-vector vibrating along the x axis. It is seen that this expression is independent of β at the viewing angle indicated.

The expressions given in the caption to Fig. 4 for the angular distribution of the photoelectrons ejected in photoionization have been shown to be applicable to electric dipole photoionization of randomly orientated molecules (as expected in most experimental arrangements), by Tully et al. [18] and Cooper and Zare [19]. The asymmetry parameter $\beta_f(E)$ may take values between +2 and −1. An atomic s electron ejected as a p-wave has a β value of 2, which is also independent of energy: but, in general, β varies with the energy of

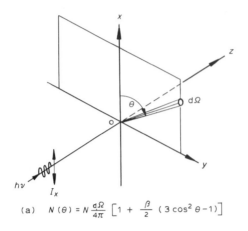

(a) $N(\theta) = N\dfrac{\mathrm{d}\Omega}{4\pi}\left[1 + \dfrac{\beta}{2}\,(3\cos^2\theta - 1)\right]$

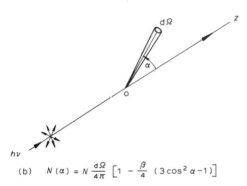

(b) $N(\alpha) = N\dfrac{\mathrm{d}\Omega}{4\pi}\left[1 - \dfrac{\beta}{4}\,(3\cos^2\alpha - 1)\right]$

Fig. 4. The angular distributions of photoelectrons released in electric dipole transitions by (a) radiation linearly polarised along Ox, and (b) unpolarised radiation. Note that when experiments are conducted to determine β using (b), the signal must usually be corrected to allow for the variation in the collision volume viewed, due to the rotation of the detector about O: $N(\alpha) \propto$ (signal at α) $\times \sin \alpha$.

the photoelectron ejected from an orbit of angular momentum l, as the radial matrix elements for the $l \pm 1$ partial waves and the phase difference between them varies [19]. Figure 5 illustrates this variation of β for the argon (3p) subshell.

In contrast, in molecular photoionization, the electronic angular

14

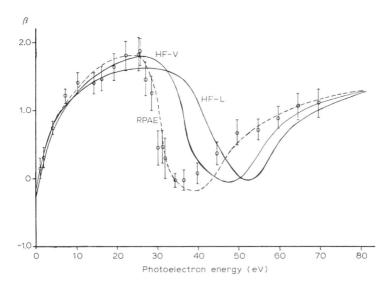

Fig. 5. The angular asymmetry parameter β for argon $3p$ electrons. Experimental points: Houlgate et al. [47]. Theory: HF—L, HF—V Hartree—Fock length and velocity calculations of Kennedy and Manson [133]. RPAE, random phase approximation with exchange calculations by Amusia et al. [181]. (From Houlgate et al. [47].)

momentum l about the centre of mass is not uniquely defined and so more than two outgoing waves are involved for ionization from any particular molecular orbital. Since, in the photoionization process, the final state is a continuum state, while the initial state is discrete, and since in the Born—Oppenheimer approximation the vibronic motion may be separated out as in Sect. 2.B, the value of β may be expected to remain constant over a photoelectron band. For σ^{-1} ionization, a large value of β (say 1—2) may be expected, while for π^{-1} ionization, β may be expected to take small positive or negative values.

Because β will vary from one molecular orbital to another, and as the photoelectron energy varies in a spectrum, it is important to consider whether or not the overall variation in β is significant in a particular investigation. Of course if the energies of the photoelectrons, and the corresponding I.P.s to the various final states are the experimental objective, any variation in β can be disregarded provided no serious loss of intensity results. This has been the situation for many studies made to date. However, as finer details of molecular structure are studied, the relative intensity of the bands in the spectrum becomes significant; and so it is becoming more important to minimise or otherwise allow for this variation in β. This is readily effected by the experimentalist making observations at the "magic" angle using, for example, a cylindrical mirror analyser (see p. 47).

The variation in β, which is an irritation in many cases, may be exploited in others to investigate the bands in a photoelectron spectrum. Since the variation of β over a particular band is rather small, overlapping or merging bands may be separated from one another with the aid of the supplementary knowledge of the β variation through the band complex [see, for example, benzene, Sect. 4.E (3)(b)].

(G) AUTOIONIZATION

In photoelectron spectroscopy, the final state is (ion in a particular state f) + (free electron with appreciable kinetic energy). A small change in the photon energy would be expected to change little but the photoelectron energy and investigations may be expected to be rather free from the rapid variations which characterise photoabsorption measurements. This freedom from resonance structure is one of the main advantages experimentalists expect to derive from the method.

However, from time to time an autoionization resonance may be "hit" by the probing photon. This situation has been discussed theoretically by Fano [20], Mies [21] and Bardsley [22], but a simple physical description will suffice here. The autoionizing state M^{**} must be reached by the usual dipole selection rules, from which it must decay to a state of the same symmetry. In our terminology, the process is

$$h\nu + M_i \rightarrow M_{f_1}^+ + e \qquad \text{direct ionization}$$

$$\rightarrow M^{**} \rightarrow M_{f_2}^+ + e \qquad \text{autoionization}$$

and the distribution over the final states populated by the decay of the autoionizing state may be very different from those excited directly. Clearly, both the Franck—Condon overlap of the M^{**} state with the initial state and the coupling between M^{**} and the final

16

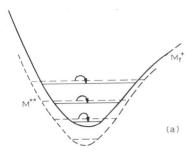

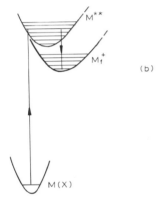

Fig. 6. Molecular autoionization. (a) Via a vibrationally excited Rydberg state
M^{**} $(n,v') \rightarrow M_f^+$ $(v'-1)$ + e. (b) Via an electronic state to a (remote) final
state.

continuum state is important. The resonance nature of the process
may be emphasised by noting that the photon energy must equal the
energy of the state M^{**} to $\hbar/\Delta t$, where Δt is the lifetime of M^{**}
against autoionization.

Often, experimentalists wish to ensure that only direct ionization occurs to the final states to minimise problems in analysis. We
may usefully distinguish two autoionizing situations (see Fig. 6). In
the first [Fig. 6(a)], the neutral state M^{**} is a vibration–rotation
excited state of a molecular ion core with an excited electron in a
high Rydberg orbital, of principal quantum number n. Autoionization then consists of a transfer of vibrational energy to the Rydberg

electron which is released, and the ionic core undergoes a relaxation. Since vibrationally excited, high Rydberg series must be expected to terminate on every final ionic state with a potential minimum, this process (see Berry [23]), which is particularly efficient for $\Delta v = -1$, can be expected to be important within a volt or two of the threshold for the various ionic states [see, for example, H_2. (Chupka and Berkowitz [24])].

A second situation is depicted in Fig. 6(b). Here, an electronically "superexcited" state, involving, for example, either the excitation of a relatively tightly bound electron or the simultaneous excitation of two electrons, decays to the final continuum state. If the intermediate and final states interact, the state M_f^+ may be strongly excited despite the fact that its vibrational levels are largely remote from the Franck—Condon region for the initial state, as shown in the figure. A good example of this type of autoionization is provided by the photoelectron spectrum of O_2 obtained using Ne resonance radiation. In this case, the $\tilde{X}^2\Pi_g$ and $\tilde{a}^2\Pi_u$ states of O_2^+ are strongly affected by the 16.85 eV, but not the weaker 16.67 eV, component of the neon resonance line. To explain the observed spectrum, at least one superexcited state of O_2 is required (see Price [25]).

The experimentalist who wishes to avoid complications due to autoionization obviously has to pick an exciting wavelength where the molecule under study is free from superexcited states. Since the number of such states must be expected to decrease above about 20 eV, the He I resonance line at 21.22 eV is often suitable. The He II resonance line at 40.8 eV is yet more suitable, but experimental problems are often associated with the use of this line. It is not always clear that published results are free from autoionizing complications, and this is particularly true if rather structureless "backgrounds" are considered as part of the spectrum, since many superexcited and final states must be expected to be totally repulsive in the Franck—Condon region. It would seem increasingly desirable for experimentalists to make investigations using dispersed radiation to check for autoionization effects (see Sect. 3.C).

It has been shown theoretically [26] and experimentally [27] that the angular asymmetry parameter β can be expected to vary sharply in the neighbourhood of an autoionization resonance, and so the variation in β may be exploited to examine autoionization. However, apparatus may have been constructed to ignore β variations by working at the "magic" angle to measure correctly the partial cross-

18

sections or branching ratios to all final states at a given wavelength. If these ratios are thought to be perturbed by autoionization, it will be necessary to investigate the possibility by changing the wavelength of the exciting radiation.

(H) PHOTOION SPECTROSCOPY

In the photoionization process, the molecular ion formed may be unstable and dissociate rather than fluoresce; most molecular ionic states do dissociate if such a channel is open. Thus we may write

$$h\nu + XY \rightarrow (XY^+)^* + e$$
$$\downarrow$$
$$X^+ + Y, X + Y^+$$

If the lifetime of the ion XY^+ is short ($\leqslant 10^{-14}$ s) then any vibrational structure in the photoelectron spectrum may be reduced and the peaks broadened, and so a study of the presence or absence of such broadening may indicate when a molecular ion undergoes dissociation (e.g. Delwiche et al. [28]). However, molecular ions frequently undergo sufficient vibrations before dissociating for the kinetic energy of the photoelectron to be well defined, while in other cases, so many final states overlap that the photoelectron spectrum does not exhibit much structure anyway. Hence the presence or absence of vibrational structure in a photoelectron band cannot be considered a reliable indication of stability, and photoion spectroscopy must be used to determine the fate of the ion over a longer time scale (typically to 10^{-7} s or so).

In general, the species of the ions formed must be determined by photoion mass spectrometry, while the final states can only be found if the kinetic energy (including any vibrational and rotational energy) of the particles formed is determined. A rather complete picture may be obtained by a photoion—photoelectron coincidence system in which the species and kinetic energy of the ions formed in a unimolecular decay are determined in coincidence with the analysed photoelectron (see Brehm and von Puttkamer [29]). The breakdown pattern of a molecular ionic state may then be traced, and an indication obtained of how the internal energy is distributed between the particles finally produced.

An analysis of the kinetic energy of ionic fragments cannot usually be made with high resolution, because the thermal energy of the parent molecular ion substantially broadens the energy of the fragments formed in the decay [30]. For example, a thermal (300 K) ion $(X_2^+)^*$ which decays to $X^+ + X$ with a kinetic energy release of 2 eV will produce ions X^+ with an energy peaked at 1 eV, but with a spread of about 0.38 eV FWHM. This energy spread may be reduced by using special sources in which the motion of the parent molecules is reduced in the direction taken by the ions whose translational energy is measured. But because of this energy spread, and because of the experimental difficulty of measuring and interpreting broad features in photoelectron spectra, kinetic analysis of photoions is particularly appropriate for the investigation of repulsive states in which the photoion energy exhibits broad peaks in any case, and good resolution is not required.

3. Experimental techniques and methods

(A) PRINCIPLES

The principal components used in ultraviolet photoelectron and photoion spectroscopy are
(a) a suitable source of radiation,
(b) a monochromator (this may be omitted in certain cases),
(c) a vacuum system into which the species may be introduced,
(d) an electron/ion analyser, and
(e) particle detector(s) and recording apparatus.
A block diagram of a typical experimental system is shown in Fig. 7.
It should be noted immediately that since all extreme ultraviolet radiation is readily absorbed, usually no window can be inserted between the source and the interaction region (but see Kinsinger et al. [31]). An excellent survey of the methods used to produce, disperse and generally handle ultraviolet radiation is provided by Samson [32].

(B) ULTRAVIOLET LIGHT SOURCES

(1) Line radiation

Atomic line radiation in the extreme ultraviolet is a convenient and frequently utilised primary source of radiation. A discharge

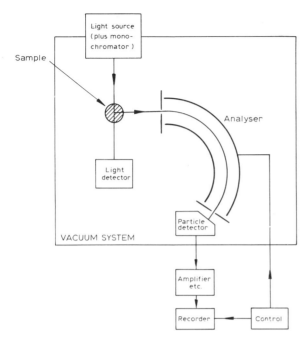

Fig. 7. The components in a typical experiment.

through helium produces copious amounts of He I resonance radiation at 58.4 nm (21.22 eV) and, under appropriate conditions, little else. The use of other gases produces a many-lined spectrum and therefore necessitates the use of a monochromator coupled to the source to select the line to be used.

If the experimentalist requires only the resonance lines of the rare gases (see Table 1) then a simple discharge tube will suffice. The discharge may be excited by the application of a steady potential to produce a glow discharge [32,33], or by using a radio frequency or microwave generator [34,186]. The former method is simple and avoids possible problems arising from electrical interference with sensitive recording systems, while the latter may prove to be more stable, although requiring a suitable coupling method to get the power into the discharge.

A discharge tube light source is shown in Fig. 8.

TABLE 1

Atomic line radiation important in photoelectron/photoion spectroscopy

Discharge conditions	Transition	Wavelength [a] (nm)	Energy [a] (eV)	Approx. intensity [b]
Pressure ~1.0 Torr	He I $2p$–$1s$	58.435	21.217	$I_0 \approx 10^{10}$–$10^{11}\ s^{-1}$
	He I $3p$–$1s$	53.705	23.086	$0.02\,I_0$
Pressure ≤0.5 Torr	He II $2p$–$1s$	30.378	40.812	$0.1 \rightarrow 0.5(I_0/4)$
(low as possible)	He I $3p$–$1s$	25.632	48.37	$0.001(I_0/4)$
	He I $(2p)^2$–$1s2p$	32.029	38.709	$0.01(I_0/4)$
Pressure ~0.5 Torr	Ne I $3s\,\{3/2\}$–$2p^6$	74.372	16.670	$0.4\,I_1$
	$3s'\{1/2\}$–$2p^6$	73.590	16.848	$I_1\ (\approx 10^{10}$–$10^{11}\ s^{-1})$
Pressure ~0.1 Torr	Ne II $2s\,2p^6(1/2)$–$2p^5(1/2)$	46.239	26.813	$0.05\,I_1$
	$2s\,2p^6(1/2)$–$2p^5(3/2)$	46.073	26.910	$0.1\,I_1$
Pressure ~0.5 Torr	Ar I $4s\{3/2\}$–$3p^6$	106.666	11.623	$0.07\,I_2$
	$4s'\{1/2\}$–$3p^6$	104.822	11.828	I_2
Impurities	O I	130.22	9.52	
	N I	113.41	10.93	
	H I (Ly α)	121.57	10.2	
Cool, Pressure ~0.5 Torr	Helium transitions	Line width $\Delta h\nu/h\nu \approx 1.5 \times 10^{-4}$ depending on conditions (see Samson [35])		

[a] Taken from Bashkin and Stoner [36] or C.E. Moore [37].

[b] All intensities are very approximate and may be affected by discharge conditions. This is particularly true of the high energy lines, and intensities are given as a guide only. Allowance must be made for the transmission of any monochromator coupled to the source, which will reduce all line intensities especially those at the shorter wavelengths. Impurity lines are obviously strongly affected by gas purity and are therefore not quoted.

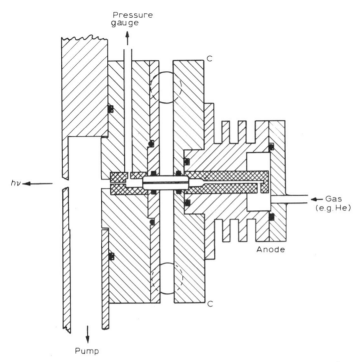

Fig. 8. Capillary discharge light source. The discharge runs through the central quartz capillary between molybdenum electrodes. Other parts in aluminium alloy or brass. Fixing screws are not indicated. For horizontal mounting, the anode is furnished with an insulated ring clamp at C. The anode is isolated using ceramic spheres, which also ensure proper alignment. For further details, see text.

(a) He I resonance radiation (2p → 1s, 58.4 nm, 21.22 eV). This and the resonance lines of the other rare gases may be produced readily with the tube shown in Fig. 8. The discharge is operated at a pressure of about 1 Torr ($\sim$130 N m^{-2} or Pa), and the discharge is maintained by a high voltage/ballast resistor combination. Typically, the high voltage, V, required to sustain the discharge is $\sim$+2 kV and the ballast resistance $R \sim$ 10 kΩ, but the values are not critical and the output of the lamp may be varied by adjusting V and/or R to increase or decrease the discharge current. A larger "starting" voltage >3 kV is generally required to initiate the discharge. If the voltage across the tube is reversed, care must be

taken that the electrons produced by the discharge do not scatter into or ionize within the interaction region. The dimensions of the capillary are not critical, but typical values are 0.1 cm diameter by 3 cm long. Cooling is generally unnecessary in this case, particularly if quartz is used for the capillary.

The provision of differential pumping and a second collimating aperture ensures that only small amounts of gas enter the main vacuum system and that the light beam is not self-reversed. To reduce the amount of impurity lines (see Table 1), it is advisable to purify the gas by using a liquid nitrogen-cooled trap containing activated carbon at the gas inlet.

(b) He II resonance radiation (2p → 1s in He$^+$, 30.4 nm, 40.8 eV). This is only produced with some difficulty by capillary discharge sources under the appropriate conditions. It is always accompanied by He I radiation, although a selective filter is available to increase the relative intensity of the He II line [60]. A high current density in a discharge at low pressure through pure helium is certainly required to produce an appreciable intensity in this line. To maintain the discharge, which tends to extinguish as the gas pressure is reduced, a large cathode area may be required, together with a power supply capable of delivering 100 mA at 6 kV or more. To operate such a discharge, it is necessary first to strike the discharge at a pressure of about 1 Torr and then reduce the gas pressure to about 0.2 Torr while increasing the voltage applied to the ballast resistor.

The design shown in Fig. 8 may be used for this purpose. The discharge capillary (0.1 cm diameter × 3 cm long) is made from quartz or boron nitride to withstand the heat generated in the

---

Fig. 9. (a). Duoplasmatron light source. The magnetic flux excited by the coils M, passes across the gap between the intermediate electrode and the anode; the stainless steel portion prevents a "short" in the magnetic circuit. The small gap between the intermediate electrode's external section and the main flange containing the anode, provides electrical insulation while completing the magnetic circuit without undue loss. The channels, w, t provide water and transformer oil cooling. The intermediate electrode has a small tungsten insert in the discharge region. For horizontal mounting, an insulated clamp at C fixes the source to the main flange. For further details, see text. (b) Section of spectral output from an argon arc in the duoplasmatron shown in (a). The light was dispersed using a Seya—Namioka f/25 monochromator, and the flux indicated refers to the light available at the exit slit. (From Fryar [40]).

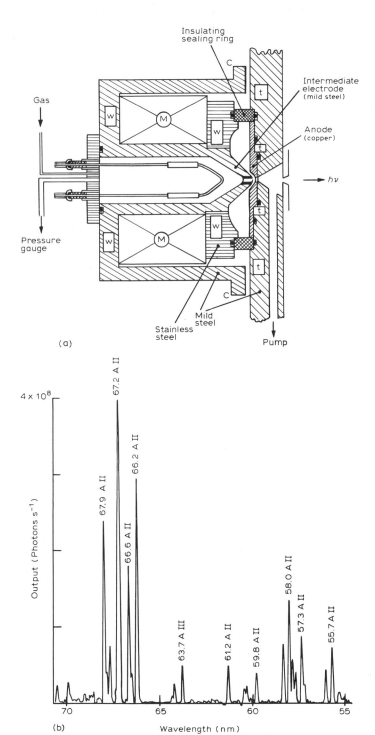

(a)

Gas

Insulating
sealing ring

C

Intermediate
electrode
(mild steel)

Anode
(copper)

w

M

w

t

t

→ hν

t

Pressure
gauge

w

M

w

t

t

C

Mild
steel

Stainless
steel

Pump

(b)

4×10^8

Output (Photons s^{-1})

67.2 A II

67.9 A II

66.2 A II

66.6 A II

63.7 A III

61.2 A II

59.8 A II

58.0 A II

57.3 A II

55.7 A II

70

65

60

55

Wavelength (nm)

discharge and maintain the gas purity, and some cooling is necessary. By furnishing the anode with cooling fins and ensuring good thermal contact in the discharge region, the problems associated with water cooling can be avoided, and forced air cooling is sufficient. It is obviously necessary to make sure that all vacuum joints are either fused, or made with Viton gaskets or by some other means to withstand the high temperatures generated near the discharge capillary.

The production of the He II line is discussed further in refs. 61 and 62.

(c) Other line radiation between 20 and 40 eV. With the exceptions noted in Table 1, only transitions between ionic states, or inner shell transitions can give rise to radiation in the wavelength range of interest. Consequently, discharge conditions have to be adjusted to encourage the production of ions, and under such conditions many spectral lines are produced from which a particular one may be selected. A complete list of available lines is out of place here, and the interested reader is referred to Samson [32] and Kelly [185]. It may be noted that many of the most prominent lines belong to the ions of the rare gases.

The excitation of some of these lines may be accomplished in a capillary discharge source under similar conditions to those described above for the production of He II resonance radiation. However, more lines at a greater intensity can be excited using the duo-plasmatron light source, a version of which is shown in Fig. 9. A small section of the spectral output from this source is also shown, which may be seen to include radiation from both singly and doubly charged ions.

The duoplasmatron is normally used as an ion source, but since the problems of producing an intense ion beam are essentially similar to the problems of producing intense spectral lines from ionised species, the arc in a duoplasmatron ion source is a very effective source of atomic line radiation. The duoplasmatron arc is maintained by electrons from a heated filament and confined by a magnetic field in the gap between the anode and intermediate electrode. The power dissipation in the source is substantial and so extensive cooling is necessary. Typical operating conditions include an arc current of 3 A at 150 V, a magnetic field of the order 0.3 T (not critical) and a gas pressure around 0.06—0.4 Torr. This light source has the advantage of producing many spectral lines, particularly in the 20—30 eV range, but a considerable price has to be paid in terms of complexity and operational convenience. Further details may be obtained from refs. 32 and 38—40.

(2) Synchrotron radiation

Since 1963 when Madden and Codling [41] first used synchrotron radiation to investigate autoionizing states in helium by photo-absorption, there has been a dramatic increase in its use. Synchrotron radiation is the name given to the electromagnetic radiation emitted by a fast charged particle orbiting in a magnetic field. Since the current circulating in an electron synchrotron may be very high and the radial acceleration substantial, a large flux of radiation can be produced by such an accelerator. This high photon flux extends usefully from ~0.1 to 200 nm and beyond in a typical machine. The electron beam is bunched and may have a height of <1 mm, and the radiation produced from the beam is both polarized and pulsed. Increasingly, storage rings are being developed which provide or will provide a suitable source for a multitude of experiments.

Because of the nature of synchrotron radiation, its use in a particular application cannot be justified unless the advantages offered by the source are adequate. Since a detailed consideration must be made in any particular instance, the following paragraphs summarise the main properties, advantages and limitations of the source. It is hoped that the properties of synchrotron radiation will be sufficiently well surveyed for the present purposes for the reader to see whether it is likely to prove suitable in a particular instance: but for more comprehensive reviews of its properties and possible applications the reader is referred to Codling [42] and a useful bibliography [43].

(a) The intensity—wavelength relation. The spectral flux available from all synchrotron sources may be expressed in a universal form, to a good approximation (Suller [44]) shown graphically in Fig. 10.

The precise spectral distribution for any machine depends on the value of a critical wavelength λ_c (=$1.86/BE^2$ nm) which in turn depends on the energy of the orbiting electrons (E in GeV) and their radial acceleration, determined by the local magnetic field (B Tesla). From the point of view of this article, however, it may be seen that the ultraviolet flux available from such a machine is approximately

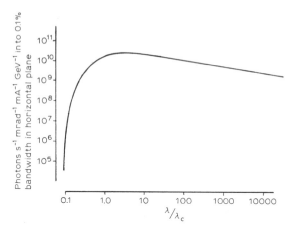

Fig. 10. Spectral output from synchrotron. The curve is "universal". The wavelength is expressed in terms of λ_c, where $\lambda_c = 0.56\ R/E^3 = 1.86/BE^2$ nm; radius R in metres, energy E in GeV, and magnetic field B in Tesla. (From Suller [44].)

constant at $\sim 10^{10}$ photons s^{-1} mA^{-1} GeV^{-1} mradian^{-1} in the horizontal plane for a 0.1% bandwidth. Of course, this figure must be reduced according to the transmission of the monochromator used to select the desired wavelength from the continuum available.

To compare this intensity with other sources, we may note that the usable intensity from a source of helium resonance radiation [45] is probably $\sim 10^{11}$ photons s^{-1}. At a comparable bandwidth, a 2 GeV storage ring with a 1000 mA circulating current coupled to a monochromator of 10% efficiency with other realistic properties would produce 10—100 times more flux [46]. This radiation is then available over the full range of photon energies of importance to the subject of this chapter. It may be noted that the monochromator efficiency will be reduced at the shorter wavelengths, but this reduction will be compensated to some extent by the larger flux available from the synchrotron (see Fig. 10).

(b) The angular divergence of the radiation. The radiation emerges with a small divergence ($\sim mc^2/E$, i.e. about $1/2E$ mrad, for an energy E in GeV) about a tangent to the orbit. However, the spread of the photon beam from its central peak is wavelength dependent. This fact may be utilised, since the flux observed at an angle of ~ 1 mrad

28

(0.6°) to the central maximum is not much reduced from its peak value for wavelengths between 10 and 100 nm. At such an angle, the X-radiation, which is most damaging to diffraction gratings in monochromators and may prove troublesome in many experiments, is virtually eliminated.

(c) The polarization of the radiation. In the plane of the orbit, the radiation from a synchrotron is 100% polarized. But it seems likely that in the circumstances considered in this chapter research would be conducted using the radiation emitted at a small angle to the tangent to the electron's orbit (see above). At such an angle, the polarization is much reduced, but it will be modified, in any case, by the polarization produced by reflections and diffraction in the monochromator used to select the wavelength of interest. A high degree of linear polarization can thus be provided at an interaction region, if the polarization produced by the monochromator increases that produced by the machine; a degree of polarization in excess of 90% may be expected in these circumstances.

This high degree of polarization may be effectively utilised to study the β asymmetry parameter and its variation with energy [47]. Its use to study the possible angular variation of the products following molecular dissociation [48] has yet to be exploited.

(d) The time dependence of the radiation. The radiation emitted by a synchrotron will, in general, consist of a series of pulses at a relatively low repetition rate (a few Hz) with a time structure to each pulse which is characteristic of the radio-frequency power used to accelerate the electrons. In a storage ring, the low frequency variation is removed because the ring is filled, and the structure remaining reflects the radio frequency used to maintain the electron energy against the radiation losses. The radiation is thus modulated with pulse length ~ 0.2 ns and spacing of ~ 2 ns. This feature of synchrotron radiation could be used for a number of studies in the field covered by the article, but to date little use seems to have been made of this property (but see Guyon et al. [49]).

(3) Other sources

A large number of variations on the sources described in Sect. (1) above can be found in the literature, and the reader should be aware that an abundance of atomic line radiation may be obtained using a condensed spark discharge, various forms of which have been described in the literature (Samson [32]). But in many ways such discharges are unsuitable for experiments of the type described in this article because of their intermittent character, variable output, and very large brightness during the period of the spark, not to mention their associated electrical noise problems.

In addition to the photon sources considered above, a simulated source of radiation is most worthy of mention. This is the application of the technique of electron scattering at small momentum transfer to simulate photons (see, for example, Brion [50], Backx and van der Wiel [51]). By measuring the energy loss of forward-scattered electrons of about 10 keV from a target species, the effective photon energy can be determined, and the equivalent photo-electron spectrum may be produced from a determination of the coincidence rate at each energy acquired by an electron ionized from the target. Various other arrangements may be made which are equivalent to optical experiments, such as photoionization mass spectrometry. The technique is particularly useful when a wide range of photon energies is required, analogous to that produced by a synchrotron, provided a modest resolution (~ 1 eV) is sufficient; but it must be pointed out that the method requires the careful use of sophisticated techniques for its effective application.

(C) MONOCHROMATORS

(1) Is dispersion required?

Many results have been published in the literature on photoelectron spectroscopy which have been taken using undispersed He I/II resonance radiation [see Sect. 3.B (1)], and in many cases this has proved sufficient. However, as the subject develops, it would seem either desirable or necessary for the light to be dispersed, so that the character of the radiation used is fully appreciated by the observer. This is particularly true when investigating the following: small peaks or structures in a spectrum; intensities of peaks, etc., when a contribution may be expected to arise from an impurity line; minor constituents in a sample; autoionization phenomena; and any other situation in which more than one process may contribute to the signal detected. However, it may be noted that in some cases a high resolution electron analyser can be used together with a knowledge of the likely spectral quality of a light source to reject or identify features in a photoelectron spectrum and so eliminate the need for dispersion.

It is vitally important for an experimentalist to decide whether or not to monochromate the light used because of the attendant disadvantages. These are principally (i) cost, both capital and running; (ii) the loss of intensity; (iii) the problem of obtaining a small spatial extent in the beam to be utilised, a problem related to cost; (iv) the adjustment and positioning of the monochromator; (v) possible interference with the particle spectrometer used, by the exit slit in particular; (vi) the variable polarization which results, and its direction with respect to the analysers and (vii) other problems such as increased constructional effort and complexity.

Despite these problems it seems certain that experimentalists will use monochromators to an increasing extent as scientific necessity is matched by improvements in monochromator performance.

(2) The selection of a monochromator

So many factors influence the selection of a monochromator in a particular application that it is not realistic to consider all the various options in relation to all the possible experiments. Nevertheless, a few general points may be made, in particular to supplement the book by Samson [32] and gather together some of the most pertinent features contained therein. Special monochromators, such as that described by Howells et al. [52] for synchrotron radiation, will not be considered here.

At the wavelengths of interest (ca. 10—100 nm), vacuum monochromators use diffraction gratings with special coatings to avoid absorption. For the same reason, a minimum number of reflections must be employed in the optical system, and because the reflection coefficient increases rapidly with angle of incidence, such reflections should be made at glancing angles. Platinum and gold are relatively good reflectors in this spectral region and do not deteriorate with time unless abused. These considerations effectively eliminate the use of the convenient Czerny—Turner and Ebert—Fastie types of monochromator, and reduce the effectiveness of the Wadsworth mount.

Due to the complexity of the equipment to be used in the

investigations considered here, a fixed exit slit and emergent beam direction are required. The light source and the analysis system beyond the exit slit are usually rather large, so that a monochromator with entrance and exit slits which are well separated from one another is also necessary. Further, low transmission loss, high resolution and small image size are highly desirable features. It is also advantageous, and sometimes essential, that the entrance slit position remains fixed too, i.e. only the grating may be moved to scan the wavelength. No instrument embodying all these features can be made, because sharp focussing cannot be maintained as the wavelength is scanned. For an instrument using a normal concave diffraction grating, the slits and the tangent at the centre of the diffraction grating must always lie along the Rowland circle, the diameter of which is equal to the radius of curvature of the grating for sharp focussing and good resolution.

The various constraints listed above severely reduce the options available to the experimentalist. At grazing incidence, one or more of the desirable features (such as fixed slit positions or fixed emergent beam direction) have to be dropped, and such instruments generally have an unsuitable wavelength coverage for the present purposes. Consequently, the experimentalist will generally select a monochromator with an off-Rowland circle mounting. The optical design of such a monochromator will then be a compromise incorporating a selection of the desirable features. The simplest and most frequently used is the Seya—Namioka mounting, named after its inventors [53,54]. An instrument of this type is available commercially [55], and is described in greater detail by Samson [32]. An asymmetrical design, incorporating different object and image distances and a high luminosity, fixed slit positions, and other features which are similar to those found in the Seya instrument, is also available commercially [56a].

In the Seya—Namioka mounting, the basic features of which are shown in Fig. 11, light from the source passes through an entrance slit S_1 to fall on the diffraction grating which is simply rotated using a sine drive to give a linear wavelength scan of the spectrum over the exit slit S_2. The monochromator is evacuated, and vacuum flap valves may be incorporated in the slit housings to isolate the vacuum chamber containing the grating from the source and detector regions.

Since the light is incident at an angle to the plane of the grating, a short line source at S_1 gives rise to an elongated image at S_2: the

32

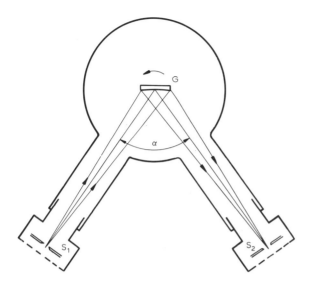

Fig. 11. Seya—Namioka monochromator. S_1 and S_2 are the entrance and exit slits, G is the diffraction grating. The monochromator is evacuated. $\alpha = 70°15'$, and the slit—grating distance is equal to $R/1.2245$ where R is the radius of curvature of the cylindrical grating.

system is inherently astigmatic. Despite this fact, careful design enables an optimum resolution of the order 0.02 nm to be obtained. However, useful intensities in the output beam are generally obtained with slit widths ~0.1 mm wide when the resolution is reduced to ~0.2 nm for a 1200 line mm^{-1} grating. Of course, the effective energy resolution in the monochromated beam is sometimes determined by the width of the line radiation dispersed, as in Fig. 9, rather than by the resolution of the monochromator.

Due to the astigmatism mentioned above, the image at S_2 obtained from a slit of length l at S_1 has a length $l + (2/3)h$ where h is the vertical height of the rulings on the grating illuminated. Realistically, the line focus at S_2 is about 2 cm long. The light beam emerging from the exit slit changes shape due to the astigmatism until it comes to a diffuse sagittal (horizontal) focus well beyond S_2. The experimentalist working beyond the exit slit who does not wish to contend with large beam dimensions must work as close as possible to the exit slit and stop down, with the inevitable loss of intensity. The astigmatism may be reduced, however, by using a concave or cylindrical foremirror placed between the entrance slit and the grating so that the astigmatism which it produces cancels that produced by the grating. Such a system is particularly useful when running costs must be reduced, because the foremirror faces the source and becomes contaminated rather than the diffraction grating, and may be recoated at a fraction of the cost of a new grating.

A better solution for reducing the astigmatism is to use an ellipsoidal or toroidal grating surface rather than the simple concave type. A toroidal grating has a reduced radius of curvature in the vertical plane compared with that in the horizontal plane in which diffraction takes place. An instrument employing a toroidal substrate with a grating surface produced holographically is now available commercially [56b]. Since the instrument is corrected for astigmatism, small straight slits are used. The mounting system is similar to the Seya, but the incident light beam strikes the grating at a more glancing angle (α =142° in Fig. 11), and consequently the short wavelength performance is much improved. With a suitable laboratory source, the makers claim that $\sim 5 \times 10^9$ photons s^{-1} of He II (30.4 nm) radiation is available. The grating is particularly expensive and so a plane foremirror would seem to be necessary, unless the instrument is used with a very clean source of radiation (such as synchrotron radiation). The resolution claimed at full aperture, f/10, is 4—5 Å with the 550 line mm^{-1} grating supplied.

An alternative approach is being developed by Speer et al. [57] and is available commercially [58]. This system operates even closer to grazing incidence and is thus suitable for the spectral range below 20 nm. A modular system is used to take advantage of grating developments as they become available; a module deploying a toroidal substrate is currently under production. It seems clear that the production of high quality dispersed ultraviolet radiation is an area where rapid technical developments can be expected [59].

Other techniques of improving the quality of far UV beams which have been described for particular purposes include the use of a filter to improve the 30.4/58.4 intensity ratio from a helium lamp [60], and a new type of transmission grating [63]. Thin metallic films may also be used to filter the light from the source and isolate the source region, but such films are very delicate and are not in general use [31].

34

(3) Light detection

Rather weak beams of ultraviolet light may be monitored easily and sensitively using the blue fluorescent light which is produced when the radiation falls onto a surface coated with sodium salicylate. A thin translucent coating may be applied as a fine mist onto a glass disc, which is then viewed using a conventional photomultiplier [64,65]. The output from such a system is proportional to the intensity and largely independent of the wavelength of the radiation over a wide range, a useful fact that may be used to make comparisons of intensities at different wavelengths (Samson and Haddad [66]).

(D) THE INTERACTION REGION

(1) Vacuum requirements

Since photoionization cross-sections are low, good vacua are not required because photon absorption must be avoided. A good vacuum is desirable or necessary, depending on the circumstances, for the successful operation of electron spectrometers, mass spectrometers and some particle multipliers. If the signals recorded are to represent the photoionization processes accurately, photoelectron or photoion peaks must vary linearly with the number densities (i.e. the pressure, at constant temperature) in the interaction region. In electron spectrometry, elastic scattering by the background gas may become evident, depending on the circumstances, at around 10^{-5} Torr. Also, "collision peaks" may be produced due to electrons suffering characteristic inelastic energy losses before they reach the analyser. No safe pressure can really be stated because so many factors are involved, such as the length of the electron trajectory, the various scattering cross-sections, the proximity of adjacent peaks, resolution required, etc. In photoion spectrometry, charge exchange and collisional dissociation processes may be particularly important near the interaction region as symmetric charge exchange and accidental resonance cross-sections are generally large at low energies. Thus it would seem sensible to build into the vacuum system a larger pumping speed and to construct it using better materials than may initially seem warranted. A very useful manual which includes much basic data with lists of equipment, performance and manufacturers is available [67] which may be used to supplement general advice on the production of high vacua such as that provided by Ward and Bunn [68].

(2) Target preparation

The sample has to be introduced into the vacuum chamber to intersect the photon beam. It is obviously preferable to be able to vary the target number density (pressure) in this region without too great an increase in the background pressure, taking note of the comments made above. A pressure gradient between the interaction region and the rest of the system may be produced quite effectively by an appropriate technique. If the sample is introduced through a wide bore tube, no useful pressure gradient can usually be maintained.

A technique suitable for volatile materials or permanent gases is to use a capillary, or a multichannel capillary array. Jones et al. [69] and Lucas [70] describe the properties and use of such systems and their methods of fabrication or purchase. Lucas recommends the use of a focussing array of small capillaries with the input pressure adjusted so that the local mean free path is less than 10% of the tube length. Selecting from his results [71] for helium the conditions required for a typical investigation, we note that he obtained an axial beam density of $\sim 2 \times 10^{19}$ m^{-3} and a beam halfwidth of ~ 2 mm about 50 mm beyond the surface of a focussing capillary array by using a throughput of $\sim 6 \times 10^{17}$ s^{-1}. This target density corresponds to a pressure of about 0.066 Pa or 5×10^{-4} Torr, and the advantage obtained by such an array may be understood by noting that the background pressure achieved using a 1 m^3 s^{-1} pump, due to the gas load, was only 0.002 Pa or 1.5×10^{-5} Torr.

A more modest increase in the target gas density may be obtained using a single capillary, for example a simple hypodermic needle. In this case, the tip of the needle should be placed as close as possible to the interaction region, and care should be taken that the gas flow is not affected by imperfections in the capillary.

Non-volatile materials may not present the same background gas problems as the material can be directed towards a cold trap to condense there. However, the material must be vaporised close to the interaction region, and the (resistive) heating coils of the miniature oven used may produce unwanted electric and magnetic fields

36

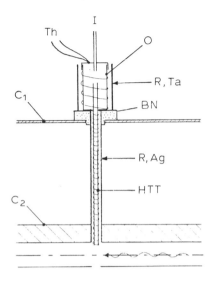

Fig. 12. Oven for high temperature vapours. C_1 and C_2 are the outer and inner cylinders of a cylindrical mirror analyser. The temperature of the platinum oven O is sensed by the Pt—Pt + 10% Rh thermocouple Th. BN is a boron nitride insulator. Heat loss from the oven O and high temperature transfer tube HTT is minimised by the tantalum and silver radiation shields R. Gases for use in calibration can be admitted through I. (From Berkowitz [45] by courtesy of the American Institute of Physics.)

besides the associated heating of other adjacent components. Systems for introducing high-temperature vapours into the rather inaccessible interaction volume of a cylindrical mirror analyser have been described by Berkowitz [72] and Potts et al. [73] (Fig. 12). Bulgin et al. have described a system which may be used to 2500 K [74].

(3) Defining the photon beam

The photon beam is required to photoionize the sample in a controlled region. Photoelectrons from surfaces contribute to the background frequently observed on photoelectron spectra and should be avoided where possible.

Figure 13 shows a system which has been devised to meet particularly stringent requirements in this regard (Strathdee and

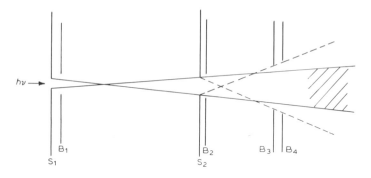

Fig. 13. Defining the photon beam. The photon beam intercepts the target in the shaded region: distances at right angles to the direction of propagation of the photon beam are greatly accelerated for clarity. Sharp slits S_1 and S_2 define the edges of the beam from the source, but inevitably act as small secondary sources. Black surfaces B_1-B_4 do not intercept the beam, but prevent photons from the slit edges or surfaces entering the region outside the dotted lines.

Browning, unpublished). The interaction region is shown shaded. The overall slit system ensures that the maximum extent of the photon beam is accurately defined, and if no apparatus intrudes into the volume through which the photons pass, an essentially zero photoelectron background is produced. It should be noted, however, that this assumes that photons do not return after passing through the interaction region. Perfect absorption is difficult to achieve in practice, but a fine black velvet surface placed well beyond the interaction region serves quite well.

(4) Defining the potential of the interaction region and the exclusion of stray photoelectrons

One of the factors governing the resolution which can be achieved with an electron analyser is the homogeneity of potential of the interaction region. This problem is considered later (p. 49) in relation to the electrostatic analysers used to analyse photoelectron energies, but a few remarks are appropriate here.

The potential of the interaction region will be defined by the surface potential of the electrodes or box surrounding it. If these are held at a fixed and constant potential, it is tempting to believe that the potential of the interaction region will be likewise fixed. Unfortunately, patch fields (see Sect. 3.E, below) and field penetration effects may produce an unwanted potential gradient across this region. The experimentalist must be particularly careful to avoid field penetration along the photon beam, since to exclude unwanted photoelectrons etc. it will probably be necessary to bias nearby slits and possibly the interaction region relative to other components. Further, to define the interaction potential, it is wise to surround the region with a single material so that contact potential gradients are avoided: but the problem of introducing the sample and allowing the photon beam to pass freely through the region makes this desirable feature impossible to achieve. Nevertheless, by careful attention to the materials and electrodes in this region, unwanted potential gradients can be avoided.

(E) ENERGY ANALYSERS

The central element in photoelectron spectroscopy is the energy analyser. Since this element is common to many other research fields in physics, including, in particular, electron impact spectroscopy, an extensive literature has arisen in which the design and performance of a wide range of energy analysers has been studied. This section is therefore written with the following aims in mind: to indicate significant factors involved in the selection of an analyser; to introduce the reader to some of the most useful types available; and to point to experimental procedures necessary for their successful operation.

(1) Selecting an analyser

In the selection of an analyser, an experimentalist should take account of the following:

(i) The range of experiments to be performed. For example, the sensitivity of the cylindrical mirror analyser is obtained at the expense of any ability to measure angular distributions, which are probably most readily investigated using a hemispherical or cylindrical 127° analyser.

(ii) Sensitivity. In photoelectron and photoion spectroscopy, the particles are generally formed in a region which is larger than the entrance aperture of the analyser. A large étendue (entrance area X entrance solid angle) is required to ensure a high luminosity and thus a good sensitivity, and Heddle [75] has calculated this parameter for the cylindrical mirror and hemispherical analysers in various modes. His results clearly demonstrate the superior sensitivity at realistic resolving powers of the cylindrical mirror, and indicate how the sensitivity of an analyser may be improved for a given resolution by retarding the energy of the electrons before analysis. The sensitivity of the hemspherical analyser should always exceed that of the cylindrical 127° analyser, since the former collects over a wide range of azimuthal angles.

(iii) Resolution. Most types of analysers appear to be capable of high resolution if carefully used, with the exception of time of flight instruments. The sensitivity at a given resolving power is the important quantity.

(iv) Other factors, such as commercial availability and price, will clearly influence the experimentalist. One factor which should not be ignored is the availability of experience. Many experimentalists appear to stick to the analyser they initially select because they appreciate its advantages and limitations, and this seems to indicate that arguments concerning the relative merits of analysers are not found convincing or compelling.

(2) Analyser types

Since a very wide variety of types of analyser exists, only a useful selection will be described here. For a broader coverage, the reader is referred to the review by Steckelmacher [76]. The electron trajectories and other associated parameters may be calculated, but these are omitted here. They may be found in the literature cited or in the article by Rudd [77]. Sufficient details are presented below, however, for the experimentalist to make an appraisal of the types selected.

Six electrostatic analysers suitable for photoelectron (photoion) spectroscopy are shown diagrammatically in Fig. 14 (i)—(vi). Electrostatic analysers are favoured over magnetic types because of the difficulties of containing and controlling the magnetic fields beyond the analysers, combined with the large deflections which electrons suffer in such fields. All the analysers shown are of the deflection type except for (i), which employs a retarding field. In the deflection analysers, particles from a source, S, follow a curved path under the influence of a deflecting field to arrive at a slit/detector system, D, which is used to separate the particles of different energy as the

40

deflection voltage, V, is scanned. The spectrum produced by such analysers is termed differential, since the analysers are designed to focus the electrons of energy E within an energy spread ΔE to the detector. The retarding analyser (i) permits all electrons with an energy greater than E to reach the detector. The spectrum thus produced is termed integral, since the flux reaching the detector is $\int_{V_R}^{\infty} N(E)dE$, where V_R is the retarding potential.

(a) Retarding field analysers. Figure 14 (i) shows an analyser which uses such a field. Analysers of this type were employed particularly in early studies of photoelectron and electron impact spectroscopy. In the design shown schematically in the figure (due to Spohr and von Puttkamer [78]), photoelectrons emerge from the source, S, to pass through a collimating lens prior to analysis by the simple retarding system shown. Two parallel meshes at potentials 0 and $-V$ retard the beam so that only particles with sufficent energy pass through the second grid and thence to the detector, D. For the successful operation of such a system, it is obviously necessary for the lens to direct a sufficiently parallel beam into the retarding field. A simpler type of retarding field analyser employs spherical grids centred on the source of the photoelectrons (Samson [79]). Good resolution can be obtained with such systems only under favourable conditions, but they can be designed with a large aperture.

It has been pointed out above that spectrometers based on this principle normally yield an integral spectrum, although by using a difference technique (the retarding potential difference or RPD technique, see Fox [80] and Chantry [81]) a differential spectrum may be produced. However, shot noise makes this technique unsuitable for the analysis of small peaks on large backgrounds (Taylor [82]).

For the successful operation of the retarding method, the retarding field must be stable and uniform. If high resolution is to be achieved, the field strength will have to change abruptly and particles will have to be reflected close to the mesh surface. Field penetration through the mesh then distorts the particle trajectories by producing local lens effects, modifying the detection efficiency, in particular of those particles which are traversing the mesh with little energy, and hence distorting the final spectrum produced. These problems may be reduced with a loss of transmission by using more meshes (Stephanakis and Bennett [83]), but the presence of the meshes

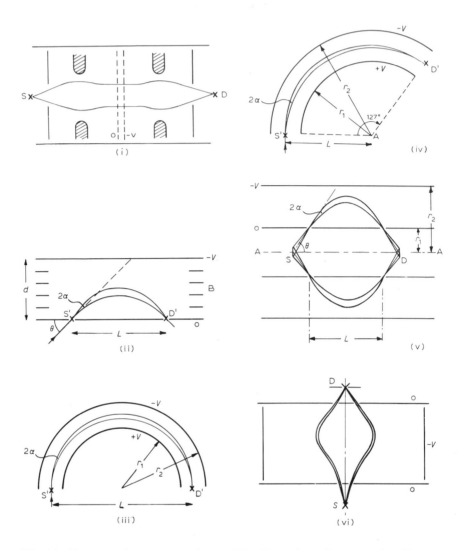

Fig. 14. Electrostatic energy analysers. The dimensions shown on the diagrams define the quantities used for the calculation of the properties of the analysers (see text). (i) Retarding field analyser (cylindrically symmetric about axis SD). (ii) Parallel plate analyser. (iii) Hemispherical analyser (spherically symmetric). (iv) 127° cylindrical analyser (cylindrically symmetric about perpendicular axis through A). (v) Cylindrical mirror (cylindrically symmetric about axis AA). (vi) Allen Bessel box [100] (cylindrically symmetric about axis SD).

42

itself limits the usefulness of the system. Any contamination may give rise to surface potential variations or patch effects which are inevitably troublesome since the particle trajectories pass through or close to the mesh surfaces. Also, the use of a longitudinal magnetic field to collimate the particle trajectories may give rise to a distorted spectrum (Simpson [84], Anderson et al. [85]).

(3) Deflection analysers: important parameters

The following general features characterise the deflection analysers shown.

The electron energy, E, (eV) passed, which is proportional to the potential difference, V volts, applied to the deflecting plates, i.e. $E = k_0 V$. The order of magnitude of k_0 is given by the ratio of the distance over which V is applied to the "characteristic dimension" of the analyser, L.

The "dispersion", D, is the displacement of the image produced by a (unit) fractional change in the electron energy. The dispersion depends on the size of the analyser, and so we may write $D = k_1 L$ where k_1 is approximately unity.

A "resolving power", ρ, may be defined as a measure of the ability of an instrument to resolve two adjacent peaks in an energy spectrum. Now, since the instruments (ii)—(v) have unit magnification, the image width would be equal to the object width if aberrations could be ignored. The best combination of peak height and resolution would then be obtained by making the object and image slit widths both equal (to some convenient value ω), and the triangular peaks produced in the spectrum would have a full width at half maximum $\Delta E_{1/2} \propto \omega$. A convenient but rather optimistic measure of the resolving power [75] is therefore

$$\rho = \frac{E}{\Delta E_{1/2}} = \frac{D}{\omega}$$

In practice, of course, the image of the object slit will be wider than ω due to the aberrations which increase rapidly with the divergence of the beam accepted by the analyser. The divergence is usually defined by the semiangle α rad., measured in the plane of deflection (see Fig. 14) and the corresponding semiangle β in the perpendicular plane. Taking account of these aberrations, the resolving power may be written

$$\rho = \frac{D}{\omega + (k_2\alpha^2 + k_3\beta^2)D}$$

The dimensionless constants k_2 and k_3 depend on the analyser used; k_2 and k_3 are both about 0.5, unless the instrument is designed to focus in the azimuthal direction, when $k_3 \sim 0$. Analysers may then be optimised by making the aberration terms approximately equal to the slit widths (Rudd [77]).

Summarising the parameters defined above, we have for each analyser

Characteristic dimension L

Working equation $E = k_0 V$

Dispersion $D = k_1 L$ $\qquad\qquad\qquad\qquad\qquad\qquad\qquad\qquad$ (8)

Resolving power $\rho = \dfrac{E}{\Delta E_{1/2}} = \dfrac{D}{\omega + (k_2\alpha^2 + k_3\beta^2)D}$

(a) The parallel plate analyser [Fig. 14 (ii)]. The dimensions are as shown in the figure, and the constants for substitution in eqns. (8) are $k_0 = 2d/L$; $k_1 = 1$; $k_2 = \frac{1}{2}$; $k_3 = \frac{1}{4}$.

The source of electrons or ions is either positioned just outside the analyser or imaged at S′ in such a way that the particles enter the deflecting field at an angle $\theta = 45°$. At such an angle, particles have their maximum range, L, and thus the parallel field causes the particles to be focussed at D′. The parallel field is simply obtained using two parallel plates, with supplementary plates B to maintain field uniformity. A practical instrument has been described in detail by Eland and Danby [86], in which the angular acceptance is limited to $\alpha \approx 1.2°$ to achieve a resolving power, ρ, of about 370 using slit widths $\omega \approx 0.18$ mm and $L = 127$ mm. Practical advantages advanced for the parallel plate analyser include the simplicity of construction, earthed entrance/exit plate, and relative freedom from patch fields. However, the angle of entry into the analyser is rather awkward: the instrument is not as sensitive as a cylindrical mirror, and is not well suited to measurements of angular distributions.

A parallel plate spectrometer, in which the particles traverse field-free regions before entering and after leaving the spectrometer at 30°, has been described by Green and Proca [87]. Second-order focussing is achieved in their design, but the instrument would be expected to be less sensitive than the 45° instrument, and problems may be posed by field penetration through the relatively wide slits. It seems probable that this design is not to be preferred for photoelectron spectroscopy.

(b) The hemispherical analyser [Fig. 14 (iii)]. The dimensions are as shown in the figure, and the constants for substitution in eqns. (8) are $k_0 = (r_2/r_1 - r_1/r_2)^{-1}$; $k_1 = 1$; $k_2 = \frac{1}{4}$; $k_3 = 0$.

The source is normally focussed to produce a divergent pencil which enters the space between two hemispherical surfaces as shown. They are then brought to a second focus at D′ and are subsequently detected. Since the spheres extend over 2π sr, focussing is achieved in the direction perpendicular to the plane shown in the figure, so that a substantial slit length and range of azimuthal angles may be utilised.

The instrument is popular not only as an electron analyser but also as an electron monochromator. Thus Kuyatt and Simpson [88], who have considered the performance of such an instrument in detail, include a discussion of effects which are not relevant to the circumstances considered here. Hemispherical analysers are particularly suited to applications where a high resolving power and directional sensitivity are required, such as in measurements of angular distribution functions. The use of a focussing retardation lens to increase the sensitivity or étendue of the instrument has been discussed by Heddle [75].

Since a radial field gradient exists at the entrance and exit slits and for other reasons, the use of full hemispheres and 180° deflection is normally combined with a lens system to transport the particles to and from the virtual slits S′ and D′. Such complexity is often not warranted in ultraviolet photoelectron and photoion spectroscopy because the energies of the particles involved are often not substantial nor very wide ranging. Simpler systems may be devised if the deflection angle is reduced somewhat since the source and detector are then placed a short distance from the field boundaries for good focussing (Purcell [89]). Real slits may then be employed in association with Herzog electrodes [90] to terminate the field near the analyser. A system following these lines has been described by Siegbahn et al. [91,92], who have selected a deflection angle of 157.5° and considered, both experimentally and theoretically, the best compromise conditions. A commercial instrument with a 150° deflection, which gives a sensitivity guaranteed at better than $5000 \ s^{-1}$ on the benzene 9.25 eV peak at a resolving power of about 300, is also available [93]. This type of system offers the additional advantage of focussing over a small plane beyond the field boundary, and so a small part of an energy spectrum may be investigated with a very high efficiency by placing a position-sensitive detector, such as a multichannel plate electron muliplier, in the focal plane.

(c) The 127° cylindrical analyser [Fig. 14 (iv)]. The dimensions are as shown in the figure, and the constants for substitution in eqns. (8) are $k_0 = 2 \ln(r_2/r_1)^{-1}$; $k_1 \sim 1$; $k_2 = \frac{1}{3}$; $k_3 = \frac{1}{4}$.

The cylindrical analyser has proved a popular and useful instrument for photoelectron spectroscopy, probably due to its simple construction and suitability when combined with the linear source of photoelectrons formed by a well collimated light beam passing through a gas. Figure 14 (iv) shows that focussing is achieved in a cylindrical sector field after an angular deflection of 127° $(\pi/\sqrt{2})$. In the design used by Turner [94], a source of photoelectrons is placed a short distance before S′ to provide an entrance solid angle of 0.0128 sr into the analyser, which is constructed so that $r_2 = 11$ cm, $r_1 = 9$ cm. Two important modifications to the simple analyser were also described by Turner [94]. Firstly, distortion of the entrance and exit fringing field was minimised by forming real slits at S′ and D′ from the small gap at the apex of two half-cylinders, whose axes coincided with the ends of the condenser plates and whose potential was held at ground (i.e. the potential at the centre of the analyser). Secondly, a thin strip electrode 1 mm wide was placed at the centre of each main electrode to modify the local field for the marginal orbits while scarcely affecting the field experienced by particles at or near the principal orbit. These electrodes were held at a potential opposite in sign to that of the main electrodes, determined empirically. Their effect was to improve the line shape in the photoelectron spectra and hence the optimum resolution. When optimised in this way, the instrument gave a resolving power of about 500.

When cylindrical analysers with simple curved plate electrodes are used as monochromators for electron impact spectroscopy, a loss of resolution and an inability to pass low velocity electrons may be experienced. Marmet and Kerwin [95] attributed these difficulties to electron reflection and space charge build-up inside the analysers.

Due, no doubt, to the much lower currents, larger geometrical size and different detailed construction of instruments used, such problems have not plagued photoelectron spectroscopists. They have, however, been able to profit from the finding of Marmet and Kerwin that electron reflection may be reduced by blackening surfaces with soot, a technique which also reduces patch fields [see Sect. (4)(a) below]. But unlike Marmet and Kerwin and others, they have not had to resort to the use of fine grids (backed by plates to remove unwanted electrons) in the place of simple curved electrodes. It is possible that such precautions may become necessary if extremely fine details are required in photoelectron spectroscopy.

(d) The cylindrical mirror, [Fig. 14 (v)]. The dimensions are as shown in the figure, and the constants for substitution in eqns. (8) satisfy $r_2 = 2.72\,r_1$; $L = 3\,r_1$; $k_0 = 1$; $k_1 = 1.5$; $k_2 \simeq 0.4$; $k_3 = 0$.

In the cylindrical mirror, particles emerging from the source S pass as a conically shaped beam through a ring-shaped slit into the space between two coaxial cylinders. They are deflected in this space to pass through a second slit in the inner cylinder to the detector. Two distinct types of analyser may be distinguished. In one type (not shown) the source S is brought to an axial focus at the detector D, which allows rather large slits and solid angles to be used with a source of modest size: the particles are injected at an angle $\theta = 42.3°$ into the deflecting field (Zashkvara et al. [96]). An alternative design is shown in Fig. 14 (v). In this design, the slit near the source is imaged back on to the surface of the inner cylinder. Only first-order focussing is available, but large signals are produced due to the high luminosity or étendue which is available in this case (Sar-El [97], Heddle [75]). Further, for the geometrical conditions given above, the focussing condition involves $\theta = 54.5°$, an angle essentially equal to the "magic" angle of $54.7°$ at which the relative heights of photo-electron peaks are independent of the angular distribution of the photoelectrons (see Sect. 2.F), and are thus fully representative of partial photoionization cross-sections. These features therefore seem to show that the surface focussing cylindrical mirror analyser is to be preferred to axial focussing instruments for photoelectron spectroscopy.

The large signals which are available because a full range of azimuthal angles is accepted, and the independence of the signal on asymmetries in the angular distribution of photoelectrons, constitute the principal advantages of the cylindrical mirror analyser. These advantages are counter-balanced by some disadvantages: the system may not be used for measurements of angular distributions (but see Sar-El's modification [97]) and the source, interaction region and detector are rather inaccessible inside the inner cylinder.

The instrument is particularly suited to measurements when small signals must be expected; thus Berkowitz [72] and his collaborators have used such an instrument to investigate a wide variety of substances requiring high temperatures for their vaporisation. A pre-retarding lens system may be used to increase the resolution of such an instrument and perform calibrations on the relative size of photo-electron peaks (Samson and Gardner [98,99]).

(e) The Allen Bessel box [Fig. 14 (vi)]. Optimised by computer programme.

A rather different approach to the problem of energy analysis is shown schematically in Fig. 14 (vi). In this design (Allen et al. [100]) electrons enter a pill-box-shaped analyser through a ring-shaped slit. The conical beam is deflected within the analyser (the electric field within the box is described by Bessel functions) to emerge from a similar slit, and thence pass to a detector at D.

Although little work has yet been published using this design, it is included here because it is available commercially [101] and it possesses some useful properties. Thus, the instrument appears to have a sensitivity equal to or in excess of that of a cylindrical mirror at the same resolving power. Like the cylindrical mirror, the entrance and exit slits are at ground potential, but access to the source and detector regions is much easier. It is also easily fabricated with accuracy, being of particularly simple construction.

Design parameters are given in ref. 100. Like the cylindrical mirror, this analyser has the disadvantage that it is not suited to measurements of angular distributions, but unlike the cylindrical mirror in the form described, it does not collect photoelectrons emitted from the source at the "magic" angle $54.7°$. The instrument may have limited application for this reason. It seems likely, how-ever, that designs using rather complex fields with computer optimization will find increasing use [102].

(4) Obtaining the specified performance from analysers

The experimentalist will be aware that like any other instrument, an energy analyser must be constructed and maintained in such a way that it is capable of performing to specification, and then operated so that its performance is not degraded.

(a) The achievement of the desired field configuration. The correct design of an analyser will ensure that the surfaces are correctly shaped, with apertures and stops set so that the beam entering the analyser has the correct size and divergence in all directions. Careful alignment of the deflecting surfaces with respect to the source and detector slits will then ensure that, *if* the potentials are correctly applied *and* unwanted fields are eliminated, the instrument will perform close to specification. To ensure that fields are correctly applied, it is necessary to consider the trajectory of particles through the analyser with care. For the experimentalist to be confident of the potential or field experienced by the particles analysed, contact potential differences arising from the use of dissimilar metals must be considered and patch fields, which arise from non-uniformities on surfaces, eliminated. Every surface, wire etc., which may influence the potential within the analyser must be considered, paying attention to field penetration through apertures. The beam being analysed should never be allowed to "see" an insulating surface, since such surfaces may acquire high surface potentials from only a small accumulated charge. To this end all surfaces facing the beam must be conducting and overlap, and must be free from irregularities likely to produce patch fields. Some experimentalists achieve this last condition by careful selection of materials (molybdenum, stainless steel, and gold plate have their advocates), which are stabilised by careful cleaning and/or baking. Others treat all their surfaces with soot, deposited using a candle flame or burning acetylene. The sooting of surfaces is cheap, and produces a surface with a relatively low electron reflection coefficient (Marmet and Kerwin [95]). Surface uniformity can be judged by eye and imperfections such as fingermarks are easily spotted.

Magnetic fields must always be removed from the analysis region. Non-magnetic materials should be selected, taking particular care with such items as "stainless steel" screws, etc. A steady field — such as that due to the Earth — will deflect electrons in particular and produce very undesirable effects. The deflection produced by the analyser will be modified, resulting in a (non-linear) error in calibration, a loss of resolution and/or a loss of signal due to the apparent misalignment of apertures. Such magnetic fields may be removed by two methods, either used separately or in combination. One is to oppose the local field, mostly due to the Earth, using suitably orientated Helmholtz coils. A review of this method can be found in the article by Rudd [77]. Alternatively, or in addition, the experimentalist may enclose all critical regions with a magnetic shield made from high permeability magnetic materials such as Mumetal [103] or Co-Netic [104] alloys. The method of construction of effective magnetic shields is discussed by the manufacturers, and in a review by Wadey [105].

Alternating fields associated with power supplies and currents through ovens, solenoids etc., must obviously be minimised. Similarly, the experimentalist must be aware of the possibility of thermoelectric e.m.f.'s and small "Ohm's law" changes in potentials when establishing or monitoring potentials in the system.

Small drifts in potentials due to changes in surface potentials seem to be unavoidable at the millivolt level over an extended period. The problem is especially acute when using substances which may condense on or modify surfaces in the analyser. The experimentalist can, however, keep a check on such changes by repeated calibration (see Sect. (5), below).

(b) Scanning and energy read-out. Most modern instruments are designed to scan in energy repeatedly, while accumulating the photoelectron/photoion signal in a multichannel analyser, for example. Such a procedure effectively averages the signals obtained over fluctuations in the light signal and sample density. It is obviously necessary to ensure that the resolution of the instrument is not degraded due to unsuitable time constants in the detection system, the response time of which should be rapid compared with the scan speed.

It is often desirable to make some potential, V volts, in the control circuit of the analyser equal to the energy of the particles passed to the detector in eV. An example is provided by the cylindrical mirror analyser in the form described above in Sect. (3)(d), for which $k_0 = 1$; i.e. the energy, E, passed is equal to the potential applied to the outer cylinder, which then becomes the ideal monitoring point for the particle energy. This "direct reading" feature may be achieved electrically in other analysers by the use of suitable external circuitry.

Both these points are discussed in greater detail by Rudd [77].

50

(5) Calibration and operating modes

There are two modes of operation commonly applied in electron spectroscopy. In one (mode A) the analyser voltage V is scanned to vary the energy of the electrons passed. The resolving power ($\rho = E/\Delta E_{1/2}$), which under conditions of proper alignment etc. is determined by the geometry of the analyser, then stays constant and the ability of the instrument to resolve adjacent peaks increases as the energy is diminished. In mode B, the analyser is optimised at a fixed energy, and the potential difference between the target region and the entrance slit to the analyser is scanned, together with any intervening electron optics. In mode B, the resolution $\Delta E_{1/2}$ is fixed unless it is degraded by the acceleration/retardation system. A constant resolution may often prove valuable in the interpretation of results, and since there are obvious advantages to the experimentalist in optimising the deflection analyser at a single energy, this mode of operation should always be considered.

(a) The calibration of the energy scale. Neither mode A nor B permits the electron energy to be determined from the voltage applied and the instrument parameters alone. Of course, if the instrument has been constructed to be direct reading (see Sect. (4)(b),

TABLE 2

Photoelectron peaks suitable for the calibration of electron analysers

Calibrating substance	Ionic state	Ionization potential I.P. (eV) (Electron energy $E = h\nu - $ I.P.)
He	$^2S_{1/2}$	25.587
Ne	$^2P_{1/2}$	21.661
	$^2P_{3/2}$	21.564
N_2	$^2\Sigma_u^+$	18.751
Ar	$^2P_{1/2}$	15.937
	$^2P_{3/2}$	15.759
Kr	$^2P_{1/2}$	14.665
	$^2P_{3/2}$	13.999
Xe	$^2P_{1/2}$	13.436
	$^2P_{3/2}$	12.130
CH_3I	$^2E_{1/2}$	10.165
	$^2E_{3/2}$	9.538

above) the error in the energy read-out on mode A may be expected to be small, but a systematic shift will be present due to unavoidable variations in contact potentials etc. within the analyser. A similar error will arise in mode B. Energy calibration is simply effected, however, by introducing an inert gas, preferably with the sample, to provide characteristic peaks in the spectrum whose positions are accurately known and which therefore serve to calibrate the energy scale. The energy scale should be linear with the voltage scanned, but this will not be true if stray fields are present or the monitoring system is imperfect, and more calibration peaks will serve to test the analyser. Interpolation may then be made between calibration peaks. Well isolated photoelectron peaks, which are particularly useful for calibration purposes, are shown in Table 2. The experimentalist will always try to obtain calibration peaks as close as possible to the spectrum under examination. The energies of characteristic lines which may be combined with these substances for calibration purposes are shown in Table 1, p. 22.

(b) Intensities. Equations (4) and (5) on p. 4 show that the relative contributions to the total photoionization cross-section from each orbital may be simply derived from the number of photoionization events, i.e. from the bands in the photoelectron spectrum. Similarly, eqn. (7) shows that the relative peak heights within a band can be directly related to the Franck—Condon factors under the appropriate conditions. Thus the intensities in photoelectron bands are important and are often required.

Unfortunately, experimental observations of peaks in photoelectron spectra can only be related accurately to the corresponding photoionization cross-sections with some difficulty. Two problems arise: the fundamental problem related to asymmetries in the angular distribution of the particles produced, and the experimental problem of the variable transmission (discrimination) of the analyser.

The first of these problems has been considered already in Sect. 2.F were the advantage of working at the "magic" angle is explained, and Sect. 3.E gives the working angle for particular instruments. The order of magnitude of the error involved in measurements of peak sizes due to such asymmetries can reach a factor of 2 if observations are made at 90° to an unpolarized beam (see Fig. 4), but much larger errors are possible if the beam is polarized. If an unpolarized beam is used, this fundamental problem is unlikely to

52

be as significant as the variable efficiency of the analyser.

An allowance for the variation in the collection efficiency of the analyser in mode A (analyser voltage scanned) may be made by noting that, since the resolving power $E/\Delta E_{1/2}$ is independent of energy, both the bandpass, and consequently, the relative efficiency increase linearly with energy. Consequently, Berkowitz and Guyon [106] recommend dividing the peak *areas* by the energy of analysis to estimate relative intensities. However, the collection efficiency of analysers in this mode may be expected to fall at low energies as the resolution becomes limited by stray fields etc. (Gardner and Samson [99]), and under such conditions Woodruff et al. [107] suggest that peak areas should be divided by the analyser bandpass.

In mode B, the energy at which the particles are analysed is fixed, and provided the input conditions to the analyser do not change, the bandwidth is also fixed. Particles are then retarded or accelerated using a lens system, and were it not for the inevitable change in the number of electrons passed to the analyser by this lens as the energy is scanned, the collection efficiency would be constant. Calibration of the collection efficiency of the analyser then amounts to a calibration of the transmission of the lens system. Methods for performing such a calibration have been described by Poole et al. [108] and Gardner and Samson [99].

The experimentalist should note that Gardner and Samson [109] have published β-independent absolute partial cross-sections for selected permanent gases, obtained using a cylindrical mirror analyser for which they calibrated the collection efficiency at a constant resolution of 45 meV. These data may be used in turn to calibrate other instruments, provided they, too, are operated at the "magic" angle.

(F) PARTICLE DETECTION

(1) Single channel multipliers

In their most usual form, all the analysers employed in photoelectron and photoion spectroscopy terminate in a single detector, used to measure the intensity of the flux of photoelectrons or photoions. Since this flux generally constitutes a current below 10^{-15} A, particle multipliers are required as detectors. Electron multipliers using discrete dynode structures (photomultipliers without a photocathode) are still used to some extent, but they are easily contaminated and lose both efficiency and gain rather rapidly on exposure to air. Continuous dynode structures made from impregnated glass such as the Bendix channeltron or Mullard channel electron multiplier can be exposed to air for long periods without degrading, have a larger gain ($\sim 10^8$) and a smaller range of output pulse size combined with a rapid rise which makes them very suitable for use in timing circuits [110]. They are unsuitable for rates in excess of 10^4 c s^{-1} and their gain drops a little above 10^3 c s^{-1}, but experimentalists generally consider themselves fortunate to obtain such signals and can, in any case, generally reduce the flux quite easily. A wide variety of types is now available including detectors fitted with cone- and slit-shaped funnels to match the output from analyser slits. The correct conditions required to match these devices to amplifiers, scalars etc., is described in literature supplied by manufacturers [111].

In the applications described here, experimentalists should note the following.

(i) It is best to count the particles detected individually, and not use the multipliers as a high gain current amplifier. This is because the amplification may vary as a function of count rate and for other reasons. Further, the statistical error involved in recording "n" particles is simply related to $n^{-1/2}$, so that this source of error is simply and accurately assessed.

(ii) Electrons are detected with an efficiency which depends on their energy. This dependence is rather rapid around 150 eV but comparatively slow around 220 eV. It is frequently possible to arrange that multipliers are suitably biased so that the electrons striking the input have a constant energy of 220 eV.

(iii) Although ions are most efficiently detected at an energy of 10—50 keV, it is not convenient to accelerate them for detection at such a high energy. But their energy spread once accelerated through 3 keV is generally small, and consequently by biasing the input cone at about —3 kV, ions initially formed with an energy of 0—10 eV are detected with nearly equal efficiency. Such a biasing condition is most readily achieved by operating the multiplier with the e.h.t. of approximately —3 kV applied at the input and connecting, say, 1 MΩ resistor to the anode to act as a load.

(iv) The potential distribution near the input affects the counting efficiency. Ray and Barnett [112] recommend that a highly

54

transparent mesh be placed across the input cone and held at the input potential to obtain a high and uniform efficiency.

(2) Position-sensitive detectors

The deployment of position-sensitive detectors to improve the rate at which data are taken by detecting the particles arriving at the focal plane of a suitably constructed analyser has been discussed by Carlson [3]. As detectors improve, such techniques will become commonplace, but to date little use seems to have been made of these devices. Their exploitation requires the use of extra electronic or computer techniques which will not be discussed here.

(G) THRESHOLD PHOTOELECTRON SPECTROSCOPY

If a continuum source of photons is available, the wavelength may be scanned so that the photoionization process

$$h\nu + M \rightarrow M_f^+ + e$$

will produce electrons of essentially zero energy as soon as the photon energy just exceeds the ionization potential to the state "f", a condition sometimes referred to as "resonance".

Techniques have been developed to observe such threshold electrons [49,113—115]. These techniques rely on the fact that very low energy electrons may be directed by very weak fields, or take a measureable and well defined time to be swept to a detector under appropriate conditions. The technique adopted by Baer et al. [113] and developed by Spohr et al. [114] is shown schematically in Fig. 15. Basically, the technique uses a simple electron collection system which discriminates strongly against electrons formed with an appreciable kinetic energy. Electrons formed with a very low energy are directed to pass through tubes with a large length-to-diameter ratio to be detected with a high efficiency, while those with an appreciable energy are likely to be intercepted by the walls of the tubes. Although a little difficulty may arise due to autoionization, the technique has been shown to work well. In this form of photoelectron spectroscopy, it is sensible to refer to resolution rather than resolving power: Spohr et al. achieved approximately 0.013 eV FWHM.

"Zero energy" electron analysers of the type described in the

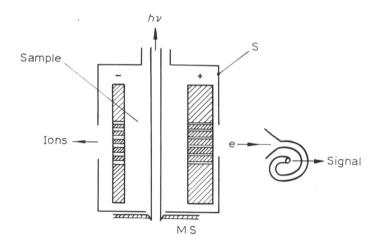

Fig. 15. Threshold photoelectron/photoion spectrometer. Very low energy electrons are directed by the uniform electrostatic field to pass through the fine channels in the positive plate. To obtain a good collection efficiency for the positive ions despite their thermal motion, the diameter-to-length ratio of the channels in the negative plate is made about 4 times larger than those in the positive plate. MS is the monochromator exit slit, and the device is surrounded by a μ-metal shield S. (From Spohr et al. [114].)

references quoted must be combined with a radiation continuum such as synchrotron radiation (Sect. 3.B(2)). The zero energy electrons are detected as the photon energy is scanned. Although autoionization processes contribute to the photoionization and, indeed, may be investigated using the same apparatus by observing the ions formed, they do not contribute to the zero energy electron signal; but since an abundance of electrons with quite a low energy may be formed by autoionization, a small unwanted signal may arise due to inadequate discrimination in the experimental system.

The zero energy electrons are detected as the photon energy is scanned, and since peaks appear as each new state of the ion is crossed, the spectrum produced is similar in appearance to a photoelectron spectrum in which the photon energy is fixed while the electron energy is scanned. However, there are some important differences. In principle, the electron energy may be better defined since their energy spread due to the thermal motion of the target molecules is eliminated. More importantly, the analysers collect electrons emitted in all directions from the target and consequently the

56

signals produced are large and essentially independent of β. The signals are proportional to the photoionization cross-section at each threshold. An important advantage of the method derives from scanning the photon energy, since an absolute calibration of energy is available without recourse to calibration using another gas. A comparative freedom from the problems associated with small drifts in contact potentials can be expected, and so it would seem that the technique could be developed to combine the advantages of conventional spectroscopic techniques with those of photoelectron spectroscopy. Further advances in the technique and its exploitation may be expected to grow with the use of synchrotron radiation (see Sect. 4.D(5), p. 81).

(H) PHOTOION MEASUREMENTS

Although most experimentalists studying the particles produced in the ultraviolet concentrate on the photoelectrons, a complete picture cannot be obtained unless the ions are also studied using one or more of the techniques described here (see also Sect. 2.H).

(1) Photoionization mass spectrometry

The ionic species may be investigated qualitatively and quantitatively using a mass spectrometer coupled to a target gas chamber into which is directed dispersed ultraviolet radiation. Quantitative measurements are only made with difficulty because the efficiency of transmission into and through a mass spectrometer is generally dependent on the direction and energy of the ions formed, so that the signal from the mass spectrometer depends on the ionic species in a way that is difficult to determine. If discrimination is not avoided, the usual consequence is that a large number of fragment ions formed with appreciable kinetic energy are lost and not detected. However, if these problems can be overcome, the ion signals for the various species formed may be combined to give the partial cross-sections appropriate to their formation, using the total photoionization cross-section which is generally obtained in a separate investigation.

Successful attempts to perform such measurements have been made by Van der Wiel and his collaborators [116] using continuum radiation simulated by high energy electrons, and in the author's

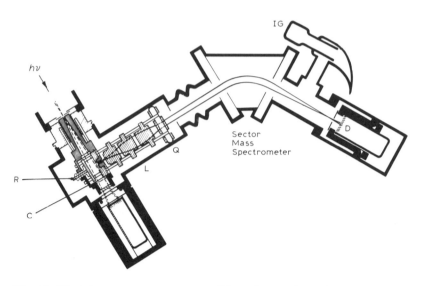

Fig. 16. Photoion mass spectrometer. Monochromatic radiation passes through an interaction chamber C biased at +11 kV. The sample is introduced into C, its pressure is monitored and the spectrometer is evacuated through ports situated above and below the plane shown. The repeller R establishes a uniform strong field in C from which photoions emerge to be focussed by the unipotential and quadrupole lenses L and Q through the sector field to the single particle detector D. The ion gauge IG indicates the pressure in the mass spectrometer, while the tapered horn eliminates spurious signals from the gauge. (From Fryar [40].)

laboratory [117]. The former system employs a specially constructed lens to focus ions with a large energy spread through a flight tube to a detector [171]. The ionic species are identified by their time of flight. The system employed by Browning et al. [117], Fig. 16, employs conventional magnetic separation of the ions. In this system, radiation from a duoplasmatron arc source is dispersed to pass through a chamber biased 11 kV above ground. A representative proportion of the photoions is swept by a uniform field in excess of 1 kV cm^{-1} through a second state of acceleration into a specially constructed wide aperture $60°$ sector mass spectrometer. An electron multiplier operates as a single particle detector and is calibrated to allow for its small (~10%) change in efficiency for various ionic species. Low pressures, typically 3×10^{-5} Torr in the target region and 3×10^{-6} Torr in the main instrument, ensure that collision processes have an insignificant effect on the measured signals. The uniform transmission characteristics of the instrument are verified by procedures which include increasing the aperture of the final detector and checking that the partial cross-sections determined can be summed to reproduce independently measured total cross-sections.

(2) Photoion energy spectroscopy

Photoion energies may be measured using the electric field analysers described in Sect. 3.E by reversing the polarity of the supplies to the analyser and operating the detector as described in Sect. 3.F.

Unless the effective thermal spread in the velocity of the species analysed is greatly reduced, by using a capillary array [Sect. 3.D(2)] for example, it is pointless operating such an analyser with a high resolution. This is because the thermal motion of the parent molecules substantially broadens the energy of the fragments formed by the dissociation of the molecular ion (Sect. 2.H).

Recognising the inevitable broadening involved in many instances, and the comparatively slow speed of travel of molecular ions compared with electrons of the same energy, experimentalists may choose to identify the mass of an ion whose kinetic energy is measured with an electrostatic analyser using a time of flight technique. In this case, the light source may be pulsed to provide a "start" signal for the ions (e.g. Eland [118]). A more accurate method of defining the start time for a photoion is to detect the associated photoelectron: such a delayed coincidence technique may even be used to measure the time of flight of H^+ ions, which travel much more quickly than heavier ions. This method may be employed in conjunction with a steady photon flux (Strathdee and Browning [119]), provided the electron detection efficiency is sufficient.

(3) Photoelectron—photoion coincidence spectroscopy

It has been pointed out already (Sect. 2.H) that a most complete analysis of the dissociation of molecular ions is made by determining the number and species of the photoions resulting from the decay of a particular ionic state identified through the photoelectron energy. The technique therefore involves a photoelectron energy analyser, the output of which provides a signal for the identification of the

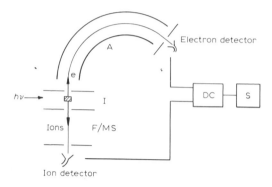

Fig. 17. Photoelectron—photoion coincidence system. A, photoelectron energy analyser. F/MS, flight tube or mass spectrometer used to determine the ionic species. DC, S, delayed coincidence/scalar system. S may be a multichannel instrument scanning synchronously with the photoelectron analyser.

species (and possibly the translational energy) of the associated photoion, which is detected in delayed coincidence. The breakdown diagram for each photoelectron energy or ionic state is thus obtained, with some indication of the translational kinetic energy of the fragment ions released. The method permits the fate of a molecular ion in a specified final state to be determined. The effort involved is, however, formidable, both in the development of the equipment and the time required for data taking.

Figure 17 shows schematically the essential features of a coincidence system. Photoelectrons and photoions are formed in a region "I" from which the photoelectrons are analysed with a photoelectron analyser "A" to provide a pulse within a small fraction of a microsecond of the initial ionizing event. The corresponding ions move under the influence of the weak extraction field in "I" through a flight tube or mass spectrometer "F/MS" to be detected some microseconds later. The photoelectron energy is defined by "A" while the ionic species is defined by the mass spectrometer, or the time of flight obtained from the delay between the electron and ion signals.

The experimental problems involved in such a system may now be appreciated. Since the photoelectron collection efficiency of a conventional analyser must be expected to be approximately 10^{-4}, the maximum signal, even if all the ions are collected and detected, cannot exceed one part in 10^4 of the potential ion signal. Signal rates

60

are limited by all the usual factors plus restrictions imposed by the resolving time in the delayed coincidence circuit. It is desirable to reduce the resolving time as much as possible, but limitations are imposed by the physical extent of the photon beam in "I" (which also limits the energy resolution in the photoelectron channel due to the potential gradient over the photoionization region) and the thermal motion or initial kinetic energy of the fragment ions formed. Background and random coincidences must be reduced as much as possible to keep this signal-to-random coincidence rate as high as possible. Since the coincidence signal is proportional to the product of the ion and electron collection efficiencies, only a minimum number of high transparency grids should be used. Final signal rates of the order of $10-10^{-2}$ s^{-1} may be expected, depending of course on the various resolutions required and the relative cross-section for the process under study.

For further details the reader is referred to the references 120—124.

4. Illustrative applications

(A) INTRODUCTION

The techniques described in Sect. 3 have been applied to the study of a very wide range of species, with the overall aim of elucidating electronic and vibrational structure, the photoionization process itself, and subsequent relaxation processes such as dissociation. The applications selected for inclusion in this section serve to illustrate some of the areas of investigation and techniques applied, but this survey cannot be considered to be comprehensive. The books and review articles listed [1—6] should be consulted to make up this deficiency.

(B) ATOMS

The photoelectron spectroscopy of atoms has been reviewed recently by Krause in ref. 4. Since atomic ionization potentials in the energy region considered here were generally well established before the application of the techniques with which we are concerned, recent investigations have concentrated on determining the role of configuration interaction or other electron correlation (EC) effects. The inclusion of EC in the description of one-electron transitions may bring calculated photoelectron energies into good agreement with experiment, but the improvement over Hartree–Fock independent single particle calculations is relatively modest. However, processes such as shake-up, double ionization and the calculation of partial cross-sections and other properties may involve EC either essentially for their description or for tolerable accuracy in their determination. Where calculations including EC are not available, its importance may be assessed from the difference between experimental values and the predictions of accurate Hartree–Fock calculations. Many of the aspects considered below are discussed rather restrictively, partly because this chapter is limited to processes which occur in the ultraviolet region of the spectrum. Inner shell processes will be discussed in another chapter of this series.

(1) Electronic states

Barium provides an excellent example of a photoelectron spectrum in which configuration mixing and other EC effects are not merely apparent but dominant. The photoelectron spectrum at 58.4 nm was investigated by Brehm and Höfler [125] and Hotop and Mahr [126] following the interesting and unexpected discovery by Brehm and Bucher [127] that the ratio of barium ions Ba^{2+}/Ba^+ formed at this wavelength is 2.4 ± 0.6.

Experimentally, the barium was produced from an oven source crossed by the photon beam and the photoelectrons from the intersection region were analysed in the usual way using a retarding field analyser of the type described in Sect. 3.E(2)(a). Extra precautions, including biassing the oven source, had to be taken to avoid detecting thermally produced electrons and to allow for contact potential differences, which both arise as the barium contaminates surfaces in the interaction region. Hotop and Mahr also used a weak electric field across the interaction region to encourage electrons ejected with a small kinetic energy towards the entrance slit of their 127° cylindrical analyser. Part of the photoelectron spectrum from Brehm and Höfler is shown in Fig. 18, taken using the retarding field analyser described by Brehm and Frey [128].

The spectrum of barium $[\ldots (5p)^6(6s)^2]$ shows some remarkable features, including (i) the population of excited states of Ba^+

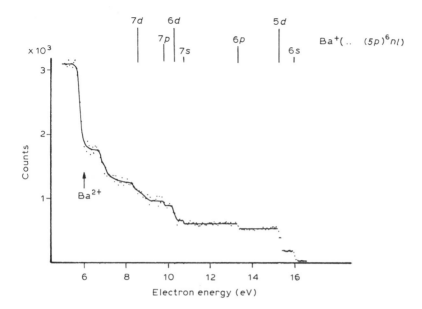

Fig. 18. Integral photoelectron spectrum of barium at 58.4 nm (21.22 eV). The thresholds for photoelectrons accompanying various excited states of Ba^+ are shown at the top of the figure: the threshold for Ba^{2+} production is also indicated. (From Brehm and Köfler [125].)

[$(5p)^6$ nl with nl = 5, 6, 7d; 7s; 6, 7p in particular] with an abundance greater than or comparable with the ground state, 6s. (ii) non-statistical branching between the fine structure levels (but see Hotop and Mahr [126] (iii) the production of Ba^+ ions in excited autoionizing states which lead to the production of Ba^{2+} ions.

Hotop and Mahr have been able to estimate lifetimes from the line widths of the electrons observed and have subsequently concluded [129] that, since the 58.4 nm line used overlaps a $(5p)^{-1}$ absorption feature (Connerade et al. [130]), an excited state of barium is formed which either decays by single electron autoionization into a number of continua to form the rich single electron spectrum mentioned in (i) above; or decays to give a two-electron ejection spectrum, with three peaks at ~0.15 eV corresponding to the formation of excited barium ions Ba^+ $(5p)^{-1}$, which subsequently autoionizes into the Ba^{2+} $(5p)^6$ continuum with the ejection of electrons in three associated peaks of about 5.85 eV. The large Ba^{2+}/Ba^+

ratio originally measured by Brehm and Bucher thus appears to derive, to a considerable extent, from a branching in favour of Ba^{2+} following the formation of the original excited state.

The interpretation of the photoelectron spectrum of barium is not yet complete, but already it can be seen to provide an interesting modern example of the interplay between various spectroscopic techniques, and demonstrates, in an extreme manner, the important role configuration mixing in the final state may play under certain circumstances.

(2) Partial cross-sections

In many cases (e.g. the rare gases), the qualitative interpretation of the ultraviolet photoelectron spectrum is trivial when compared with that of barium, quoted above; single peaks arise corresponding to each final ionic state, where these states are simply described by single electrons removed from the ground state configuration of the atom. For example, in the photon range considered here, the photo-ionization of argon [... $(3s)^2(3p)^6$] will give rise to three peaks, one ($^1S_0$) corresponding to $(3s)^{-1}$ and two others, closely spaced, corresponding to $(3p)^{-1}$, with $^2P_{3/2,1/2}$ ionic states. From this point of view, all electron correlation (EC) effects are comparatively unimportant. However, the branching ratios or partial cross-sections into the final states are influenced by EC, particularly when a weak channel opens up while a neighbouring channel is responding strongly. Fano [131] interprets such effects in terms of the polarizability of the electron shells combined with intershell coupling.

Figure 19 shows the partial cross-section for argon $(3s)^{-1}$ photo-ionization as determined by Houlgate et al. [132]. In this investigation, synchrotron radiation was combined with photoelectron spectroscopy to investigate the partial photoionization cross-sections $(3s)^{-1}$, $(3p)^{-1}$ as a function of wavelength over a wide energy range. The independently measured total cross-section was broken down by measuring the relative magnitudes of the photoelectron peaks at each wavelength. Corrections were made for the response of the 127° cylindrical analyser used as a function of the photoelectron energy analysed, for each peak, coupled with an appropriate allowance made for the bandpass of the monochromator. In this case, measurements were not made at the "magic" angle (Sect. 2.F) and so an allowance was also required for the variation of the asymmetry

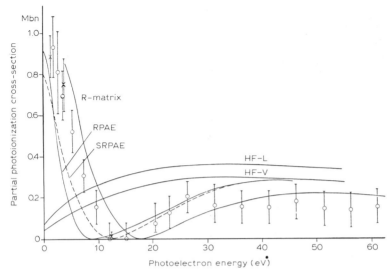

Fig. 19. Partial photoionization cross-section for argon 3s subshell. Experimental points: O, Houlgate et al. [47,132]. x, Samson and Gardner [180]. Theory: HF—L, HF—V, Hartree—Fock length and velocity calculations of Kennedy and Manson [133]; RPAE, random phase approximation with exchange by Amusia et al. [183]; R-matrix, calculations of Burke and Taylor [182]; and SRPAE, simplified RPAE of Lin [134].

parameters $\beta(E)$ appropriate to the peaks, and for the degree of polarization of the radiation used.

The resulting partial cross-section (Fig. 19) illustrates the influence of intershell coupling, as may be appreciated by comparing the experimental results with theory. The Hartree—Fock calculations of Kennedy and Manson [133] give quite an incorrect description of the threshold behaviour in contrast to the SRPAE (simplified random phase approximation with exchange) calculations of Lin [134], which include EC through intershell coupling.

(3) Angular distributions

The rare gases — and here we may quote argon again — provide examples where EC effects change the directional properties of the ejected photoelectrons. This is particularly true in the vicinity of a "Cooper minimum" (Fano and Cooper [135]). As shown in Fig. 5 in Sect. 2.F, the β asymmetry parameter for the argon 3p subshell is particularly susceptible to the photon energy between about 20 and 40 eV above threshold, where the corresponding partial cross-section is small.

Experimentally, the asymmetry parameter was measured by Houlgate et al. [47] using synchrotron radiation dispersed by a monochromator and therefore polarized. A 127° cylindrical analyser was rotated about the beam as an axis through angles up to ±100° with respect to the polarization vector. Such large rotations cannot be achieved if β-asymmetry parameters are measured using an un-polarized beam, and potentially greater accuracy can be achieved using polarized light to measure β. The β-asymmetry parameter was calculated from the relative measurements of the flux of photo-electrons obtained at each angle θ to the principal (polarization) axis ($\theta = 0$), using the formula

$$\frac{N(\theta)}{N(0)} = \frac{\left\{ 1 - \dfrac{\beta}{2} + \dfrac{3\beta}{2(g + 1)}\ (\cos^2\theta + g\sin^2\theta) \right\}}{\left(1 - \dfrac{\beta}{2} + \dfrac{3\beta}{2(g + 1)} \right)}$$

The polarization factor g (defined as I_{minor}/I_{major}) was sufficiently small in some of these measurements for its value to be rather un-important, although it was determined. For some measurements the polarization was not so complete and so g was required accurately. The value of g was conveniently determined in a subsidiary series of measurements with helium, for which it can be safely assumed that $\beta = 2$, corresponding to the single channel $(1s)^2 \rightarrow 1s\epsilon p$, in the photon range considered here.

The result shown in Fig. 5 on p. 15 shows [compare (2), above] that EC effects may have to be included for an accurate description of the angular dependence of the ejected photoelectrons; the β parameter is sensitive to values of radial matrix elements and their relative phases, and hence to their precise values at a given energy.

(C) PHOTOELECTRON SPECTRA OF MOLECULES

(1) Introduction

The principle features of photoelectron spectra have been described in Sect. 2, from which it will be appreciated that photo-

66

electron spectroscopy provides particularly valuable information for the elucidation of molecular structure. Indeed, because the technique continues to play such an important role, its application to the valence shells of molecules has been extensively reviewed. Price [2,136] has concentrated on basic concepts and relatively small molecules, while Jonathan et al. [137], Frost et al. [138] and Berkowitz [2] have examined substances requiring special techniques for their investigation. Inorganic (Dekock) and organic molecules (Heilbronner and Maier) have also be reviewed [2]. A particularly useful overview, covering a wide range of different processes and techniques, may be found in the book by Carlson [3] while Eland's monograph [1] supplements a full discussion of photoelectron spectra with a chapter on the positive ions formed as a result of photoionization, a comparatively neglected field. Extensive lists of references and summaries of the wide variety of species investigated by photoelectron spectroscopy, particularly using helium resonance radiation, are to be found in the references quoted above, in the book by Turner et al. [6] and in the articles by Betteridge et al. [139—141]. A large number of interesting papers covering various problems in photoelectron spectroscopy are contained in ref. 142, devoted entirely to this subject. In this connection, the reader must also be aware that, since 1972, a journal [143] has been largely devoted to studies of the kind referred to in this article.

In the following section, one or two examples have been selected from this extensive literature to illustrate significant results, or techniques which appear to the author to offer good prospects for future studies.

(2) Energies and the identification of molecular orbitals

The nature of molecular orbitals may be determined from the ionization potentials, photoelectron peak structure, photoelectron angular distributions and relative peak sizes, from substitution effects and by making comparisons with other similar molecules and theoretical calculations.

Since the ionization potentials in a molecule cannot be calculated accurately, matching up the photoelectron peaks with calculations is not necessarily straightforward. Full calculations including electron correlation effects have only been made for the simplest system, and even Hartree—Fock calculations are rarely available. In principle, it is not adequate merely to have orbital calculations, but a calculation in which the energy change on ionization includes the energy change resulting from the reorganization of the molecule as the photoelectron escapes. Since such calculations are often too complex, it is generally necessary to make comparisons with self-consistent-field (SCF), or Hückel molecular orbital, or CNDO, calculations of orbital energies [144] and then assume that Koopmans' theorem is valid, i.e. that the ionization potential from a particular orbital is the orbital energy with the sign reversed. An example of this procedure is given below (benzene). Calculations may sometimes be made of peak structures to aid orbital identification, and here the alkali halides provide a good example.

(3) Examples

(a) Nitrogen. An illustration of the problems inherent in comparing I.P.s determined experimentally with theoretical calculations was provided relatively early in the development of the technique of photoelectron spectroscopy by N_2. In N_2, as in isoelectronic CO, the ground state of the molecular ion is $^2\Sigma_g(3\sigma_g)^{-1}$ while the first excited state corresponds to $^2\Pi_u$, $(1\pi_u)^{-1}$. But extensive Hartree—Fock calculations by Cade et al. [145] produced a reverse ordering for these two states. Agreement between theory and experiment was only obtained when Verhaegen et al. [146] incorporated the electron correlation energy differences between these states into the calculations. Evidently, even "exact" Hartree—Fock calculations must be treated with caution.

(b) Benzene. A second, and perhaps more typical, illustration is provided by benzene, C_6H_6. We consider the orbitals responsible for its well-known structure, ignoring the C(1s) electrons. The 58.4 mm (He I) photoelectron spectrum is shown in Fig. 20 together with measurements of the associated angular distribution parameter β. The measurements of β were made by Carlson and Anderson [147] by measuring the angular variation in the photoelectron signal as noted in Fig. 4(b), taking proper account of the variation in the ionization volume involved. Also shown in Fig. 20 are the band assignments due to Jonsson and Lindholm [148], which were corroborated later by extensive calculations (von Niessen et al. [184]). Photoelectron spectra taken with higher photon energies, including the 30.4 nm

68

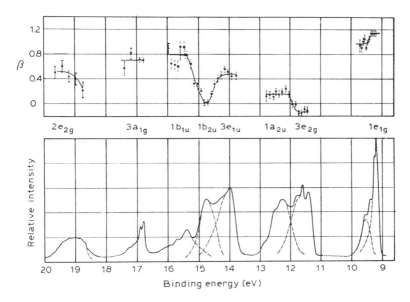

Fig. 20. The 58.4 nm photoelectron spectrum of benzene and the corresponding angular distribution parameters β. The eight principal bands correspond to the ionization of the molecular orbitals designated, following the original assignment of Jonsson and Lindholm [148] confirmed by von Niessen et al. [184]. (From Carlson and Anderson [147], by courtesy of North-Holland Publishing Company.)

He II line, have enabled Åsbrink et al. [149] to show that two more photoelectron peaks are situated at 22.5 and 28.8 eV. Now the total number of valence orbitals is 10 and, in the absence of significant configuration mixing effects, 10 distinct peaks may be expected in the spectrum. Due to band overlapping, complex contours such as that shown between 11.4 and 12 eV which arises from Jahn—Teller effects, and the approximate nature of the theoretical calculations of the orbital energies, orbital assignment is difficult. While all the calculations such as those of Schulman and Moskowitz [150] show, and experimental measurements of the vibrational structure confirm, that the first band with the lowest I.P. corresponds to the removal of a π C—C bonding orbital orientated perpendicular to the molecular plane, subsequent orbitals are not well delineated. But in 1970, Åsbrink et al. [149] confirmed the assignments of Jonsson and Lindholm, supplementing their publication of the full spectrum with an analysis of the peak areas, which they associated with the number of electrons in each peak. Then, by identifying the peaks with the highest I.P.'s with the aid of the SCF calculations, they were able to indicate that attempts which had been made to assign more molecular orbitals to the lower energy region could not be reconciled with their observations; there are simply not enough electrons in the molecule to go round. Their final assignment of the orbitals within overlapping bands is particularly acceptable, taking account of the clear breaks in the angular distribution function β determined later by Carlson and Anderson. It is quite probable that many of the problems involved initially in orbital identification would have been reduced, had such measurements been available earlier. More generally, however, it should be noted that measurements of β cannot be relied on to provide extra information if, for example, bands with a similar β overlap extensively.

(c) The alkali halides. A particularly good illustration of the necessity of performing good calculations for comparison with carefully performed experiments is provided by recent investigations of bonding in the alkali halides, which we will designate as M^+X^- in recognition of their ionic quality.

On the experimental side, relatively high temperatures are required for the vaporisation of these compounds and the experimentalist is inevitably faced with working with a low sample density. A high electron collection efficiency is thus required, and here the cylindrical mirror may be used effectively, as described by Berkowitz [72] and in Sect. 3.E. Inevitably, the sample condenses on sensitive regions of the system and impairs the performance of the electron analyser. Repeated calibrations are necessary for accuracy, using calibration gases admitted simultaneously into the vacuum system. Another particularly important experimental problem is that dimers M_2X_2 are produced by the oven, which give rise to photoelectron peaks close to the monomers under investigation. Dimer production may be discouraged (with a loss of signal) by designing an oven and beam transfer system operating at a rather low temperature. A most elegant solution to this and other similar contamination problems is to employ a photoelectron—photoion coincidence system, but this technique had not been used in this context. Nevertheless, the various experimental problems have been tackled to good effect by Berkowitz et al. [151] and Potts et al. [152]. Their results, but not the details of their interpretation, are in generally good accord and two examples are shown in Fig. 21.

It is natural to focus attention in the first place on the peaks to be found near the first ionization threshold in the alkali halides, since these correspond to the ionization in the sub-shell of the "transferred" electron responsible for the ionic bond. Higher peaks may be described in terms of the corresponding atomic ions [187]. Thus the discussion of the lower peaks essentially rests on the states of the system which may be represented as M^+X^0. Two factors are then particularly relevant, as the ionization of the X^- ions may be represented as $p^6 \to p^5$. Firstly, the halogen "atom" left would be ^2P if isolated, and so its atomic $P_{3/2}-P_{1/2}$ fine structure splitting must be expected to show up in the spectrum. Secondly, the axial field in the molecule will be expected to split the atomic p-orbital into π and σ components, according to the projection of the angular momentum along the internuclear axis. In the halogen hydrides, this splitting is

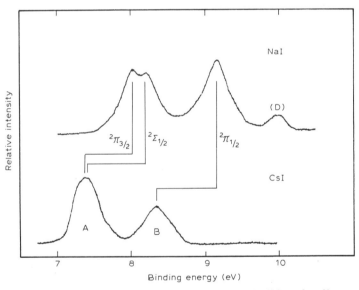

Fig. 21. Part of photoelectron spectrum of the iodides of sodium and caesium at 58.4 nm. The band assignments shown were deduced by Berkowitz et al. [2,151]. Caesium iodide spectrum from ref. 151 and sodium iodide from Potts et al. [152]. The $P_{3/2}-P_{1/2}$ splitting in free iodine is 0.94 eV. The peak D is due to the dimer $(NaI)_2$.

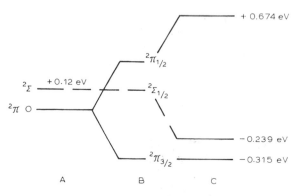

Fig. 22. Splitting diagram for CsI deduced by Berkowitz et al. [151]. A, σ—π splitting; B, including spin—orbit interaction in Π state; C, including Ω—Ω interaction.

known to be very large (approximately 4 eV in HI, for example), but Berkowitz et al. [151], quoting and extending the calculations of Matcha [153], point out that the splitting should be much smaller in the alkali halides. Their approximate calculations gave ~0.12 eV for CsI and ~0.5 eV for NaI. Since these values are smaller but of the same order as the spin—orbit splitting, the $\Pi_{1/2}$ and $\Sigma_{1/2}$ components would be expected to lie quite close to one another if they did not interact.

Berkowitz et al. then calculated the repulsion between the $\Pi_{1/2}$—$\Sigma_{1/2}$ states arising from the Ω—Ω configuration interaction, which, in the case of CsI, gives the result shown in Fig. 22. This diagram shows that, in this case, the $\Sigma_{1/2}$—$\Pi_{3/2}$ energy difference will not be sufficient to be detected in the experiment reported here, whereas the $\Pi_{1/2}$—$\Pi_{3/2}$ splitting will be large, and not greatly different from the $P_{1/2}$—$P_{3/2}$ splitting in the iodine atom. In other words, the iodine atom is nearly spherically symmetrical. Peak A in Fig. 21 thus corresponds to the composite peak ($\Sigma_{1/2}$ + $\Pi_{3/2}$) while B corresponds to the $\Pi_{1/2}$ fine structure level. In favour of this interpretation is the double peak structure for peak A shown for NaI, as measured by Potts et al. In this case, Berkowitz et al. show that a larger Π—Σ splitting is indeed to be expected, as noted above.

An alternative approach to the explanation of the peaks observed in the alkali halides has been advanced by Potts et al., in which the $P_{3/2}$—$P_{1/2}$ atomic states of the halogen atom are split by the Stark

72

field on the nearby M^+ ion. Fully developed, their approach would presumably match that outlined here, due to Berkowitz et al.; but the molecular orbital description of the lowest states of the molecular ions seems to be well supported by theoretical calculations and the best experimental work available (see also Berkowitz in ref. 2).

(4) Vibrational structure and bonding

The general factors affecting the appearance of photoelectron bands have already been stated (Sect. 2.D). It has also been noted that since the vibrational frequency in a given state is related to the bond strength, a comparison of the vibrational frequency displayed by a photoelectron band with the corresponding frequency in the neutral molecule will indicate the bonding power of the electron released.

Since any change in bonding may be expected to give rise to an associated change in molecular geometry, it is natural to suppose that the profile of the photoelectron band can be related to this change, at least empirically. Now a simple characterisation of the profile of a photoelectron band may be made by measuring the difference between the vertical and adiabatic ionization potentials, $(I_V - I_A)$. Turner [6,154] has shown that the difference $(I_V - I_A)$ is approximately linearly related to the fractional change in the vibrational frequency between the ground and the ionic states, $(\omega'' - \omega')/\omega'$. Eland [1] gives a useful semiclassical formula

$$(l\,\delta\theta)^2 = (\Delta r)^2 = 0.0054\,\frac{(I_V - I_A)}{\mu(\omega')^2}$$

where I_V, I_A are in eV, μ is the reduced mass appropriate to the vibrational mode in atomic units, ω' is in cm^{-1}, l and Δr are measured in nm, and $\delta\theta$ in radians. Δr is the change in bond length between molecular ion and neutral when a stretching occurs, while $\delta\theta$ is the change in the equilibrium bond angle in a bond of length l when bending vibrations are excited. Eland quotes examples to show that this simple formula should be good to 10% or so, when compared with the more elaborate method of Heilbronner et al. [155].

The vibrational levels excited in photoionization follow the general selection rules, given by Herzberg [156], for example. The vibrational sequences actually measured in a particular instance, however, may be correlated with the vibrational modes and hence with the bonding properties of the electron released, at least in favourable circumstances. For example, in H_2O and D_2O, in the extensive second band the bending mode ν_2 is excited strongly, while its frequency is reduced. The difference between the vertical and the adiabatic I.Ps is approx. 0.96 eV, and calculation then shows that the HOH bond angle opens up from 105° in the ground state of the neutral to ~173° in the $(3a_1)^{-1}$ ion. This is consistent with the bonding character of the $3a_1$ orbital and the fact that its removal substantially reduces the electron density between the hydrogen atoms, allowing their nuclei to repel one another and excite the vibrational sequence observed (Brundle and Turner [157], Potts and Price [158]).

(5) Franck—Condon factors, peak envelopes and the vertical I.P.

As noted in Sect. 2.B, the relative heights of the vibrational peaks observed in photoelectron bands may be compared with the Franck—Condon factors [eqn. (7), p. 5]. However, such a comparison can only be made if (i) on the experimental side, the band shape is correctly measured so that a proper allowance is made for the electron transmission variation of the analyser used, including any (small) β variation through the band; and (ii) on the theoretical side, the electronic transition moment $G(r,E)$ is constant over the band and autoionization is unimportant.

An experimental test of the variation in $G(r,E)$ over a band may be made by comparing the relative heights of the vibration peaks obtained with an appropriately calibrated analyser with the calculated Franck—Condon factors, assigning any observed differences to the variation in $G(r,E)$. Calculations of $G(r,E)$ are not generally available for comparison with such "experimental" results, however. A notable exception is the H_2^+ ground state for which Berkowitz and Spohr [159] first made careful measurements and compared various selected measurements with the calculations of Itikawa [160]. The isotopic forms HD and D_2 have also been investigated. Further calculations have since been made by Ford et al. [161], and the most recent experimental and theoretical results for H_2 are compared in Fig. 23.

This result is important in that a substantial variation in $G(r,E)$ is seen to occur across the band using 21.22 eV photons; the envelope is "distorted" by this variation. Should this variation be quite

74

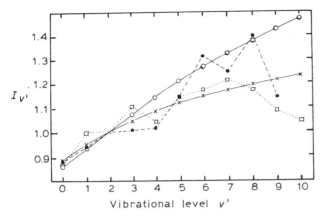

Fig. 23. Transition probabilities to v' levels of H_2^+ at 58.4 nm compared with Franck—Condon approximation. In this diagram, the relative transition probability is

$$I_{v'} = \frac{\sigma(v', 0)}{\sigma(2, 0)} \Bigg/ \frac{q(v', 0)}{q(2, 0)}$$

where $\sigma(v',0)$ is the cross-section for a transition to H_2^+ (v') from H_2 $(v'' = 0)$ and $q(v',0)$ is the corresponding Franck—Condon overlap integral for H_2^+ (v') with H_2 $(v'' = 0)$. Ratios are taken relative to the $v' = 2$, the strongest peak in the vibrational sequence in H_2^+ [see Fig. 3(a)]. For Frank—Condon transitions, $I_{v'} = 1$ throughout, so the graph illustrates the variation in the electronic transition moment $G(r,E)$ over the band. Experiments: ●- - - - - -●, Berkowitz and Spohr [159]; □· · · · · ·□, Gardner and Samson [162]. Calculations: ×————×, Ford et al. [161]; ○————○, Itikawa [160].

commonplace, particularly amongst species with unresolved bands where no check on the envelope can be made, discrepancies may be expected between the vertical I.P.s measured and those obtained theoretically. The equilibrium internuclear separations deduced for the ionic states will be affected in a similar way. Any such errors would be expected to be most significant for bands giving rise to electrons of low energy, since it is for such bands that the variation in $G(r,E)$ would be expected to be largest and the electron transmission characteristics of the analyser the most difficult to ascertain (see Sect. 3.E(5)(b), p. 52).

The dramatic change in the vibrational envelope of the $\widetilde{X}$ and $\widetilde{a}$ states of O_2^+ produced when oxygen is irradiated by the neon resonance line at 16.85 eV (see Sect. 2.G and ref. 25) is the result of autoionization. Such effects may occur less obviously and un-wittingly in other cases. Procedures such as the making of comparisons between different species known to have similar electronic structure, and changing the wavelength of the exciting radiation would generally allow the experimentalist to guard against such effects.

(6) Other techniques for comparison with photoelectron spectroscopy

Reference has already been made in examples given above to the measurements of I.P.s, the angular asymmetry parameter β, vibrational modes and spacings, the use of deuterium substitution, and the investigation of band intensities in the assignment of photoelectron bands.

Other methods which may be used are (i) comparisons with Rydberg series determined by photoabsorption (e.g. Lindholm [163]), (ii) substitution, particularly of hydrogen by fluorine (e.g. Brundle et al. [164]), (iii) examining the approximate range of I.P.s for particular molecular orbitals (Betteridge et al. [165]) and (iv) measuring the relative intensities of bands excited at different wavelengths (Price et al. [166]). Comparison may also be made with the results of Penning ionization by metastable atoms (e.g. Cermak [167]) and electron energy-loss spectroscopy (e.g. Lassettre and Skerbele [168]), particularly using the coincidence systems of the type described by Hamnett et al. [169], Brion [50], and Backx and Van der Wiel [51], in which small angle electron scattering is used to simulate photoionization as mentioned in Sect. 3.B(3), p. 30.

(D) PHOTOION MEASUREMENTS

(1) Introduction

The majority of publications analysing the particles produced by ultraviolet light are concerned with the photoelectrons. This is natural, since, as eqn. (2) indicates, it is the electron which indicates the energy transfer in the photoionization process and provides most of the information required to characterise the final ionic states. However, as noted in Sect. 2.H, the dissociation of the molecular ions formed must usually be traced by methods other than photo-electron spectroscopy. This topic is treated by Eland [1] and is

76

currently under review [2, Vol. II/III], and these references should be consulted, in particular for the theoretical basis of the subject.

Photoion measurements may be applied to answer questions such as (i) how many ions of each species are formed as a function of wavelength, (ii) how much kinetic energy is acquired by the particles formed and (iii) how the internal energy of a particular ionic state is partitioned amongst the particles eventually produced, following dissociation. Experimentally, these questions are tackled by mass spectrometry, photoion kinetic energy spectroscopy and photo-electron—photoion coincidence spectroscopy.

(2) Photoionization mass spectrometry

The nature and the relative abundance of the ions formed in photoionization may be determined as a function of wavelength, by coupling a suitable source/monochromator combination to a mass spectrometer. In practice, accurate quantitative measurements are not easily achieved, because the dissociated species are generally formed with appreciable kinetic energy, which leads to an underestimation of the dissociation products and a consequent overestimation of the near-thermal ions. Of the very few attempts to combat this problem (see Sect. 3.H) two results are shown in Fig. 24.

(a) Example, hydrogen. The results of Backx et al. [170] were taken with an electron impact system to simulate synchrotron radiation, albeit with only modest resolution (1 eV), by detecting the photoions in coincidence with forward-scattered electrons whose energy loss indicated the appropriate photon energy [50]. Photoions were identified by their time of flight. The system permits the determination of dipole oscillator strengths (proportional to the photoionization cross-section) as well as the relative numbers of H^+ and H_2^+ ions formed [171]. The results of Browning and Fryar [172] were taken with the system described in Sect. 3.H(1), p. 58.

The main features of these results can be summarised as follows. From the threshold for the production of H^+ (18.1 eV) to ca. 26 eV, dissociation proceeds directly through the H_2^+ $\tilde{X}(1s\,\sigma_g)$ state. While the variation in the branching ratio with energy may be understood in terms of Franck—Condon transitions to the repulsive part of the $1s\sigma_g$ state above the dissociation limit, good agreement between theory and experiment can only be obtained by calculations in which

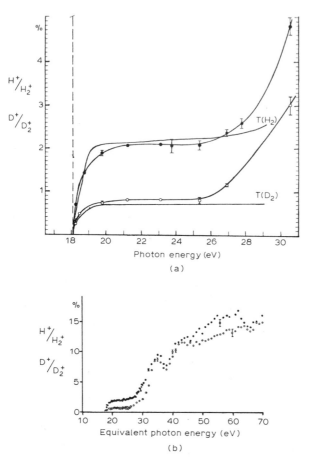

Fig. 24. The dissociative photoionization of hydrogen and deuterium. (a) From threshold of dissociative ionization to about 26 eV, through H_2^+ ($1s\sigma_g$). Experiment: ●, ○, Browning and Fryar [172]. Theory (T): Ford et al. [161]. (b) From threshold to 70 eV. Experiment: ●, ○, Backx et al. [170] (by courtesy of The Institute of Physics).

both initial and final states are accurately represented. The link-up between the vibrational envelope $I_{v'}$ (Fig. 23) and the dissociation fraction H^+/H_2^+ can now be regarded as satisfactory, both experimentally and theoretically. Above 26 eV, the situation is still far from clear. The rapid rise in the fraction of protons produced below

78

30 eV appears to be due to autoionization of doubly excited states of H_2 converging on H_2^+ ($2p\sigma_u$) (Strathdee and Browning [119], Hazi [173]). The dip at about 37 eV may be due to two factors: the loss of resonance channels producing ions, together with some neutral production through a $^1\Pi_u$ resonance state producing neutrals, $H(2s) + H(2p)$ (Misakian and Zorn [174]). At yet higher photon energies, the ($2p\sigma_u$) and higher resonance and non-resonance states are involved to a degree not yet determined.

(b) Other areas of application. As studies at shorter wavelengths become possible using new experimental techniques such as synchrotron radiation, and in particular species requiring high temperatures for their vaporization are investigated, photoionization mass spectrometry will be required to supplement photoelectron spectroscopy not only to follow dissociation patterns but also to discriminate between species and impurities present in low density beams, and to assist in the identification of final states through the associated manifolds of Rydberg levels. An important development in this area is the use of isotopic labelling to identify species and hence trace the routes whereby dissociation occurs. Berkowitz and Eland [175] have used isotopically labelled N_2O, $N^{14}N^{15}O$, to study the formation of ions, including, surprisingly, $N^{14}O^+$, in a high resolution experiment capable of examining the autoionization of Rydberg series converging on various final states of N_2O^+ [cf. Fig. 6(a)].

(3) Photoion kinetic energy spectroscopy

It has already been noted that when using a "static" gas target, the measurement of photoion kinetic energies formed by dissociation is limited in resolution by the broadening brought about by the thermal motion of the parent molecules. Some reduction in the thermal motion may be produced by measuring the ions which emerge sideways from a directed beam, produced by the techniques described in Sect. 3.D(2). But in addition, electrostatic energy analysers are incapable of distinguishing the species of ion measured without elaboration and so photoion energy measurements are often of limited application. However, the very fact that there is little to be gained by working at high resolution means that time of flight techniques may be successfully employed. Ion flight times between an ionization region and a detector may be determined using either a pulsed source (Eland [176], Guyon et al. [49]) or by using the associated photoelectron (e.g. Strathdee and Browning [119]) to define the moment of ionization, and then the pulse from the ion detector to fix the transit time of the ion.

(a) Examples, hydrogen and oxygen. Hydrogen and oxygen provide examples of these techniques in which there is obviously no ambiguity concerning the ionic species detected. In H_2 [119], the presence of energetic protons (2—4.4 eV) at photon energies below any direct channel capable of producing such energetic protons is a direct proof of the involvement in photoionization of autoionizing states subject to dissociation, first highlighted by the electron impact work of Crowe and McConkey [177]. In O_2^+, the predissociation of $B^2\Sigma_g^-$ state has been correlated with the associated photoelectron spectra by Gardner and Samson [188]. They have been able to assign most of the peaks in their observed ion kinetic energy spectra to the predissociation of the states observed in the photoelectron spectra, on the assumption that certain dissociative limits (acceptable by the Wigner—Wittmer rules) were involved while others were not. For instance, although the broad $^2\Pi_u$ state in O_2^+ centered at ca. 24 eV (5.3 eV above the O^+ (^4S) + $O(^3P)$ limit) shows no vibrational structure in the photoelectron spectrum and therefore dissociates very rapidly, the photoion kinetic energy spectrum at 23.7 eV displays no ions with a kinetic energy in excess of 1.5 eV. Gardner and Samson therefore assign the dissociation limit for the $^2\Pi_u$ state to be $O(^3P) + O^+(^3D^0)$ at 22.06 eV, giving rise to O^+ ions with an energy of approximately 0.8 eV.

(4) Photoelectron—photoion coincidence spectroscopy

In this elegant technique, mass selected photoions are detected in coincidence with energy-selected photoelectrons after ionization by monochromatic light. Within the resolution limits of the electron analyser, the energy transferred in the ionization process is defined (by $h\nu - E$) and thus the dissociation of molecular ions can be examined with a known and variable internal energy. Such experiments are technically very difficult [see Sect. 3.H(3)].

(a) Example, SO_2. Brehm and his co-workers [120—121] have

80

studied a number of species using this technique and their results on the predissociation of SO_2^+ ions [124] may be taken as illustrative. The first two bands in the photoelectron spectrum lie below the dissociative ionization limits and thus produce SO_2^+ ions only, while the third band lies entirely above the lowest dissociation limit and might be expected to produce SO_2^+, SO^+ and S^+ ions. In common with other species, it is found that in this band dissociation is complete so that no SO_2^+ ions are formed, while the ratio $SO^+ : S^+$ ions produced was found to be 75 : 1. Now, both thresholds for the production of SO^+ and S^+ ions lie below the threshold for the third band. Although the SO^+ threshold is the lower, Brehm et al. show that an application of the quasi-equilibrium theory (QET) of mass spectra would yield an SO^+/S^+ ratio $<5 : 1$, clearly at varinace with the observation. An investigation of the SO^+ translational kinetic energy was also made from the time of flight spectrum, and found to represent approximately 50% of the available energy; and, further, the translational energy spectrum of SO^+ was found to be rather broad, and not to increase significantly with excitation energy. None of these observations fits a direct dissociation model, which would predict that more than 90% of the available energy should appear as translational kinetic energy and that the spectrum of ion kinetic energies should be narrow. However, the observations of the spectrum of SO^+ kinetic energies were also in conflict with QET, which predicts a maximum intensity near zero energy. These measurements thus show that neither a statistical model nor a direct dissociation model can predict the dissociation pattern of SO_2^+. This example demonstrates how photoelectron—photoion coincidence techniques provide rather complete experimental information which any new theoretical models will have to explain.

(5) Threshold photoelectron—photoion coincidence spectroscopy

Guyon et al. [49] have embarked on a programme of particular promise. By combining the discrimination technique described in Sect. 3.G with the rather precise timing information available using synchrotron radiation, they identify threshold electrons (electrons of essentially zero kinetic energy). The associated ions are detected in delayed coincidence. As the photon energy is varied, autoionization effects can be exploited to excite levels and hence gain spectroscopic information not available in conventional photoelectron spectroscopy. Similarly, ion kinetic energies can be measured to provide the same type of information as that described immediately above, but again more states can be made accessible through autoionization. A similar technique has been described by Stockbauer [178].

(6) Other techniques

As mentioned in the introduction, photodetachment from negative ions is not considered here, and for the same reason the spectroscopy of ions involving photons in the visible or near visible regions of the spectrum has not been considered. But as tunable lasers become available in the ultraviolet, the spectroscopy of both positive and negative ions will make a dramatic entry into the field, in particular because it will become possible to investigate ionic states with millielectron volt resolution or better. This is because fast ion beams with a very low relative energy spread can be allowed to interact with the laser beam (Huber et al. [179]), and the study of photon-ion interactions can therefore be expected to provide further detailed information on molecular structure.

General references

1 J.H.D. Eland, Photoelectron Spectroscopy, Butterworths, London, 1974.
2 C.R. Brundle and A.D. Baker (Ed.), Electron Spectroscopy: Theory, Techniques and Applications, Academic Press, London, 1977, Vol. I. (Volumes II and III to be published.)
3 T.A. Carlson, Photoelectron and Auger Spectroscopy, Plenum Press, New York, 1975.
4 F.J. Wuilleumier (Ed.), Photoionization and Other Probes of Many-Electron Interactions, Plenum Press, New York, 1976.
5 J.A.R. Samson, Photoionization of Atoms and Molecules, Phys. Rep., 28C (1976) 304.
6 D.W. Turner, C. Baker, A.D. Baker and C.R. Brundle, Molecular Photoelectron Spectroscopy, Wiley-Interscience, London, 1970.

Other references

7 F.I. Vilesov, B.L. Kurbatov and A.N. Terenin, Sov. Phys. Dokl., 6 (1961) 490.
8 D.W. Turner and M.I. Al-Joboury, J. Chem. Soc., (1963) 5141. (See also ref. 6.)

9   A.L. Stewart, Adv. At. Mol. Phys., 3 (1967) 1.
10 G. Herzberg, Spectra of Diatomic Molecules, Van Nostrand Reinhold Co., Litton Educational Publishing Inc., 1950.
11 O. Edqvist, E. Lindholm, L.E. Selin and L. Åsbrink, Phys. Scr., 1 (1970) 25.
12 E. Heilbronner and J.P. Maier, in C.R. Brundle and A.D. Baker (Eds.), Electron Spectroscopy: Theory, Techniques and Applications, Academic Press, London, 1977, Vol. I.
13 B.S. Schneider and A.L. Smith, in D.A. Shirley (Ed.), Electron Spectroscopy, North-Holland, Amsterdam, 1972, p. 335.
14 J.H.D. Eland, Photoelectron Spectroscopy, Butterworths, London, 1974, p. 132.
15 A.W. Potts and W.C. Price, Proc. R. Soc. Lond., Ser. A, 326 (1972) 165.
16 W.C. Price, A.W. Potts and D.G. Streets, in D.A. Shirley (Ed.), Electron Spectroscopy, North-Holland, Amsterdam, 1972, p. 187.
17 J.A.R. Samson, J. Opt. Soc. Am., 59 (1969) 356; Phil. Trans. R. Soc. London, Ser. A, 268, (1970) 141.
18 J.C. Tully, R.S. Berry and B.J. Dalton, Phys. Rev., 1976 (1968) 95.
19 J. Cooper and R.N. Zare, J. Chem. Phys., 48 (1968) 942.
20 U. Fano, Phys. Rev., 124 (1961) 1866.
21 F.H. Mies, Phys. Rev., 175 (1968) 164.
22 J.N. Bardsley, Chem. Phys. Lett., 2 (1968) 329.
23 R.S. Berry, J. Chem. Phys., 45 (1966) 1228.
24 W.A. Chupka and J. Berkowitz, J. Chem. Phys., 51 (1969) 4244.
25 W.C. Price, in C.R. Brundle and A.D. Baker (Eds.), Electron Spectroscopy: Theory, Techniques and Applications, Academic Press, London, 1977, Vol. I, Sect. IVD, p. 189.
26 D. Dill, Phys. Rev. Sect. A, 7 (1973) 1976.
27 J.A.R. Samson and J.L. Gardner, Phys. Rev. Lett., 31 (1973) 1327.
28 J. Delwiche, P. Natalis, J. Momigny and J.E. Collin, J. Electron Spectrosc. Relat. Phenom., 1 (1972) 219.
29 B. Brehm and E. von Puttkamer, Z. Naturforsch. Teil A, 22 (1967) 8.
30 P.J. Chantry and G.J. Schultz, Phys. Rev. Lett., 12 (1964) 449.
31 J.A. Kinsinger, W.L. Stebbings, R.A. Valenzi and J.W. Taylor, Anal. Chem., 44 (1972) 773.
32 J.A.R. Samson, Techniques of Ultraviolet Spectroscopy, Wiley, New York, 1967.
33 C.R. Brundle, Appl. Spectrosc., 25 (1971) 7.
34 F.C. Fehsenfeld, K.M. Evenson and H.P. Broida, Rev. Sci. Instrum., 36 (1965) 294.
35 J.A.R. Samson, Rev. Sci. Instrum., 40 (1969) 1174.
36 S. Bashkin and J.O. Stoner, Atomic Energy Levels and Grotrian Diagrams, Vol. I, North-Holland, Amsterdam, 1975.
37 C.E. Moore, Atomic Energy Levels, N.B.S. Circular 467, 1949.
38 M.V. Ardenne, Tabellen der Electronen Physik, Ionen Physik und Ubermikroskopie, Deutsch. Ver. der Wissenschaften, Berlin, Vol. I, 1956.
39 M. Bruck, R. Bruckmüller and W. Weissman, Extended Abstracts V Int. Conf. on V.U.V. Radiation Physics, 1977, Montpelier, France, Vol. 3, p. 60.

40 J. Fryar, Ph.D. thesis, Queen's University Belfast, 1972.
41 R.P. Madden and K. Codling, Phys. Rev. Lett., 10 (1963) 516.
42 K. Codling, Rep. Prog. Phys., 36 (1973) 541.
43 G.V. Marr, I.H. Munro and J.C.C. Sharp, Report DNPL/R24, 1972 and DL/TM 127, 1974, Daresbury Laboratory, Warrington, WA4 4AD.
44 V.P. Suller, Report DL/TM 118, 1973, Daresbury Laboratory, Warrington, WA4 4AD.
45 J. Berkowitz, J. Chem. Phys., 56 (1972) 2766.
46 The Scientific Case for Research with Synchrotron Radiation, Report DL/SRF/R3, 1975, Daresbury Laboratory, Warrington, WA4 4AD.
47 R.G. Houlgate, J.B. West, K. Codling and G.V. Marr, J. Electron Spectrosc. Relat. Phenom., 9 (1976) 205.
48 G. Dunn, Phys. Rev. Lett., 8 (1962) 62.
49 P.M. Guyon, T. Baer, L.F.A. Ferreira, I. Nenner, A. Tabche-Fouhaile, R. Botter and T.R. Govers, Abstracts X I.C.P.E.A.C., 1977 (Commissariat à l'Énergie Atomique, Paris), Vol. 1, p. 96.
50 C.E. Brion, Radiat. Res., 64 (1975) 37.
51 C. Backx and M.J. Van der Wiel, J. Phys. B, 8 (1975) 3020.
52 M.R. Howells, D. Norman, G.P. Williams and J.B. West, J. Phys. E, 11 (1978) 199.
53 M. Seya, Sci. Light (Tokyo), 7 (1958) 23.
54 T. Namioka, J. Opt. Soc. Am., 49 (1959) 951.
55 McPherson Model 235 Scanning Monochromator, McPherson Instrument Corporation, Acton, Massachusetts.
56 (a) Model ASM 50 M0.02, Jobin-Yvon, 91160 Longjumeau, France.
   (b) Model LHT 30 Monochromator, Jobin-Yvon, 91160 Longjumeau, France.
57 R.J. Speer, R.L. Johnson and D. Turner, Extended Abstracts V Int. Conf. on V.U.V. Radiation Physics, 1977, Montpelier, France, Vol. 3, p. 26.
58 Grating Measurements Ltd., 57 Cornwall Gardens, London SW7 4BE.
59 Abstracts, V Int. Conference on V.U.V. Radiation Physics, Montpelier, France, 1977, Vol. 3.
60 A.W. Potts, T.A. Williams and W.C. Price, Discuss. Faraday Soc., 54 (1973) 104.
61 R.T. Poole, J. Liesegang, R.C.G. Leckey and J.G. Jenkin, J. Electron Spectrosc. Relat. Phenom., 5 (1974) 773.
62 F. Burger and J.P. Maier, J. Electron Spectrosc. Relat. Phenom., 5 (1974) 783.
63 S.A. Flodstrom and R.Z. Bachrach, Rev. Sci. Instrum., 47 (1976) 1464.
64 M. Seya and F. Masuda, Sci. Light (Tokyo), 12 (1963) 9.
65 R.A. Knapp, Appl. Opt., 2 (1963) 1334.
66 J.A.R. Samson and G.N. Haddad, J. Opt. Soc. Am., 64 (1974) 1346.
67 L. Holland, W. Steckelmacher and J. Yarwood, Vacuum Manual, E. and F.N. Spon, London, 1974.
68 L. Ward and J.P. Bunn, Introduction to the Theory and Practice of High Vacuum Technology, Butterworths, London, 1967.
69 R.H. Jones, D.R. Orlander and V.R. Kruger, J. Appl. Phys., 40 (1969) 4641.

70 C.B. Lucas, Vacuum, 23 (1972) 395.
71 C.B. Lucas, J. Phys. E, 6 (1973) 992.
72 J. Berkowitz, J. Chem. Phys., 56 (1972) 2766.
73 A.W. Potts, T.A. Williams and W.C. Price, Proc. IV Int. Conf. on V.U.V. Radiation Physics, Pergamon, 1974, p. 162.
74 D. Bulgin, J. Dyke, F. Goodfellow, N. Jonathan, E. Lee and A. Morris, J. Electron Spectrosc. Relat. Phenom., 12 (1977) 67.
75 D.W.O. Heddle, J. Phys. E, 4 (1971) 589.
76 W. Steckelmacher, J. Phys. E, 6 (1973) 1061.
77 M.E. Rudd, in K.D. Sévier (Ed.), Low Energy Electron Spectrometry, Wiley-Interscience, New York, 1972.
78 R. Spohr and E. von Puttkamer, Z. Naturforsch. Teil A, 22 (1967) 705.
79 J.A.R. Samson, Phil. Trans. R. Soc. London, Ser. A, 268 (1970) 141.
80 R.E. Fox, W.M. Hickman, D.J. Grove and T. Kjeldaas, Rev. Sci. Instrum., 26 (1955) 1101.
81 P.J. Chantry, Rev. Sci. Instrum., 40 (1969) 884.
82 N.J. Taylor, Rev. Sci. Instrum., 40 (1969) 792.
83 S. Stephanakis and W.H. Bennett, Rev. Sci. Instrum., 39 (1968) 1714.
84 J.A. Simpson, Rev. Sci. Instrum., 32 (1961) 1283.
85 N. Anderson, P.P. Eggleton and R.G.W. Keesing, Rev. Sci. Instrum., 38 (1967) 924.
86 J.D.H. Eland and C.J. Danby, J. Phys. E, 1 (1968) 406.
87 T.S. Green and G.A. Proca, Rev. Sci. Instrum., 41 (1970) 1778.
88 C.E. Kuyatt and J.A. Simpson, Rev. Sci. Instrum., 38 (1967) 103.
89 E.M. Purcell, Phys. Rev., 54 (1938) 818.
90 R. Herzog, Z. Phys., 97 (1935) 596; and Phys. Z., 41 (1940) 18.
91 K. Siegbahn, C. Nordling, A. Fahlman, R. Nordberg, K. Hamrin, J. Hedman, G. Kohansson, T. Bergmark, S.E. Karlsson, I. Lindgren and B. Lindberg, Nova Acta Regiae Soc. Sci. Ups. Ser. IV, 20 (1967).
92 U. Gelius, E. Basilier, S. Svensson, T. Bergmark and K. Siegbahn, Univ. Uppsala Inst. Phys. Rep. No. U.U.I.P. 817, 1973; B. Wannberg, U. Gelius and K. Siegbahn, Univ. Uppsala Inst. Phys. Rep. No. U.U.I.P. 818, 1973.
93 Model UVG3 from V.G. Scientific Ltd., The Birches Industrial Estate, East Grinstead, Sussex, England.
94 D.W. Turner, Proc. Roy. Soc. London, Ser. A, 307 (1968) 15.
95 P. Marmet and L. Kerwin, Can. J. Phys., 38 (1960) 787.
96 V.V. Zashkvara, M.I. Korsunkii and O.S. Kosmachev, Sov. Phys. Tech. Phys., 11 (1966) 96.
97 H.Z. Sar-el, Rev. Sci. Instrum., 38 (1967) 1210.
98 J.L. Gardner and J.A.R. Samson, J. Electron Spectrosc. Relat. Phenom., 2 (1973) 267.
99 J.L. Gardner and J.A.R. Samson, J. Electron Spectrosc. Relat. Phenom., 6 (1975) 53.
100 J.D. Allen, J.P. Wolfe, G.K. Schweitzer and W.E. Deeds, J. Electron Spectrosc. Relat. Phenom., 8 (1976) 395.
101 Model 560 Photoelectron Spectrometer, Special Instruments Laboratory Inc., 312 West Vine Av., Knoxville, Tennessee, 37901.

102 B. Wannberg, G. Engdahl and A. Skollermo, J. Electron Spectrosc. Relat. Phenom., 9 (1976) 111.
103 Telcon Metals Ltd., Manor Royal, Crawley, Sussex, England.
104 Magnetic Shield Division, Perfection Mica Co., Chicago, Illinois.
105 W.G. Wadey, Rev. Sci. Instrum., 27 (1956) 910.
106 J. Berkowitz and P.M. Guyon, Int. J. Mass Spectrom. Ion Phys., 6 (1971) 302.
107 P.R. Woodruff, L. Torop and J.B. West, J. Electron Spectrosc. Relat. Phenom., 12 (1977) 133.
108 R.T. Poole, R.C.G. Leckey, J. Liesegang and J.G. Jenkin, J. Phys. E, 6 (1973) 226.
109 J.L. Gardner and J.A.R. Samson, J. Electron Spectrosc. Relat. Phenom., 8 (1976) 469.
110 Acta Electron., 14 (2) (1971) contains a number of useful papers on these detectors.
111 e.g. Mullard Technical Information Report 16, 1975. From Mullard House, Torrington Place, London WC1E 7HD.
112 J. Ray and C.F. Barnett, I.E.E.E. Trans. Nucl. Sci., 17 (1970) 44.
113 T. Baer, W.B. Peatman and E.W. Schlag, Chem. Phys. Lett., 4 (1969) 243.
114 R. Spohr, P.M. Guyon, W.A. Chupka and J. Berkowitz, Rev. Sci. Instrum., 42 (1971) 1872.
115 S. Cvejanovic and F.H. Read, J. Phys. B, 7 (1974) 1180.
116 C. Backx, R.R. Tol, G.R. Wight and M.J. Van der Wiel, J. Phys. B, 8 (1975) 2050, 3007.
117 R. Browning, J. Fryar and R. Cunningham, Adv. Mass Spectrom., 6 (1974) 933.
118 J.H.D. Eland, Int. J. Mass Spectrom. Ion Phys., 9 (1972) 397.
119 S. Strathdee and R. Browning, J. Phys. B, 9 (1976) L505.
120 B. Brehm and E. von Puttkamer, Adv. Mass Spectrom., 4 (1968) 591.
121 B. Brehm, V. Fuchs and P. Kebarle, Int. J. Mass Spectrom. Ion Phys., 6 (1971) 279.
122 J.H.D. Eland, Int. J. Mass Spectrom. Ion Phys., 8 (1972) 143.
123 C.J. Danby and J.H.D. Eland, Int. J. Mass Spectrom. Ion Phys., 8 (1972) 153.
124 B. Brehm, J.H.D. Eland, R. Frey and A. Küstler, Int. J. Mass Spectrom. Ion Phys., 12 (1973) 197.
125 B. Brehm and K. Höfler, Int. J. Mass Spectrom. Ion Phys., 17 (1975) 371.
126 H. Hotop and D. Mahr, J. Phys. B, 8 (1975) L301.
127 B. Brehm and A. Bucher, Int. J. Mass Spectrom. Ion Phys., 15 (1974) 463.
128 B. Brehm and R. Frey, Z. Naturforsch. Teil A, 26 (1971) 523.
129 K. Gerard, H. Hotop and D. Mahr, Adv. Mass Spectrom., 7 (1976).
130 J.P. Connerade, M.W.D. Mansfield, K. Thimm and D. Tracy, in E.E. Koch R. Haensel and C. Kunz (Eds.), Vac. U.V. Radiation Physics, Pergamon, 1974, p. 245.
131 U. Fano, Comments At. Mol. Phys., 4 (1975) 119.
132 R.G. Houlgate, J.B. West, K. Codling and G.V. Marr, J. Phys. B, 7 (1974) L470.

133 D.J. Kennedy and S.T. Manson, Phys. Rev. A, 5 (1972) 227.

134 C.D. Lin, Phys. Rev. A, 9 (1974) 171.

135 U. Fano and J.W. Cooper, Rev. Mod. Phys., 40 (1968) 441.

136 W.C. Price, Adv. At. Mol. Process., 10 (1974) 131.

137 N. Jonathan, A. Morris, M. Okuda, K.J. Ross and D.J. Smith, Discuss. Faraday Soc., 54 (1972) 48.

138 D.C. Frost, S.T. Lee, C.A. McDowell and N.P.C. Westwood, J. Electron. Spectrosc. Relat. Phenom., 12 (1977) 95.

139 D. Betteridge and A.D. Baker, Anal. Chem., 42 (1970) 43A.

140 D. Betteridge, Anal. Chem., 44 (1972) 100R.

141 D. Betteridge and M.A. Williams, Anal. Chem., 46 (1974) 125R.

142 Discuss. Faraday Soc., 54 (1972).

143 Journal of Electron Spectroscopy and Related Phenomena, Elsevier, Amsterdam.

144 P.A. Cox, S. Evans, A.F. Orchard, N.V. Richardson and P.J. Roberts, Discuss. Faraday Soc., 54 (1972) 26.

145 P.E. Cade, K.D. Sales and A.C. Wahl, J. Chem. Phys., 44 (1966) 1973.

146 G. Verhaegen, W.G. Richards and C.M. Moser, J. Chem. Phys., 47 (1967) 2995.

147 T.A. Carlson and C.P. Anderson, Chem. Phys. Lett., 10 (1971) 561.

148 B.O. Jonsson and E. Lindholm, Ark. Fys., 39 (1969) 65.

149 L. Åsbrink, O. Edqvist, E. Lindholm and L.E. Selin, Chem. Phys. Lett., 5 (1970) 192.

150 J.M. Schulman and J.W. Moskowitz, J. Chem. Phys., 47 (1967) 3491.

151 J. Berkowitz, J.L. Dehmer and T.E.M. Walker, J. Chem. Phys., 59 (1973) 3645.

152 A.W. Potts, T.A. Williams and W.C. Price, Proc. Roy. Soc. London, Ser. A, 341 (1974) 147.

153 R.L. Matcha, Compendium of Alkali Halide Wave Functions, University of Houston, Texas, 1970.

154 D.W. Turner, Phil. Trans. R. Soc. London, Ser. A, 268 (1970) 7.

155 E. Heilbronner, K.A. Muzkat and J. Schaublin, Helv. Chim. Acta, 54 (1971) 58.

156 G. Herzberg, Electronic Spectra and Electronic Structure of Polyatomic Molecules, Van Nostrand Reinhold Co., 1966.

157 C.R. Brundle and D.W. Turner, Proc. R. Soc. London, Ser. A, 307 (1968) 27.

158 A.W. Potts and W.C. Price, Proc. R. Soc. London, Ser. A, 326 (1972) 181.

159 J. Berkowitz and R. Spohr, J. Electron Spectrosc. Relat. Phenom., 2 (1973) 143.

160 Y. Itikawa, J. Electron Spectrosc. Relat. Phenom., 2 (1973) 125.

161 A.L. Ford, K.K. Docken and A. Dalgarno, Astrophys. J., 195 (1975) 819.

162 J.L. Gardner and J.A.R. Samson, J. Electron Spectrosc. Relat. Phenom., 8 (1976) 123.

163 E. Lindholm, Ark. Fys., 40 (1969) 97 and following papers.

164 C.R. Brundle, M.B. Robin, N.A. Kuebler and H. Basch, J. Am. Chem. Soc., 94 (1972) 1451.

165 D. Betteridge, M. Thompson, A.D. Baker and N.R. Kemp, Anal. Chem., 44 (1972) 2005.

166 W.C. Price, A.W. Potts and D.G. Streets in D.A. Shirley (Ed.), Electron Spectroscopy North-Holland, 1972, p. 187.
167 V. Čermak, J. Electron Spectrosc. Relat. Phenom., 9 (1976) 419.
168 E.N. Lassettre and A. Skerbele, in D. Williams (Ed.), Methods of Experimental Physics, Vol. 3B, Academic Press, New York, 1974.
169 A. Hamnett, W. Stoll, G. Branton, C.E. Brion and M.J. Van der Wiel, J. Phys. B, 9 (1976) 945.
170 C. Backx, G.R. Wight and M.J. Van der Wiel, J. Phys. B, 9 (1976) 315.
171 C. Backx, G.R. Wight, R.R. Tol and M.J. Van der Wiel, J. Phys. B, 8 (1975) 3007.
172 R. Browning and J. Fryar, J. Phys. B, 6 (1973) 364.
173 A. Hazi, Chem. Phys. Lett., 25 (1974) 259.
174 M. Misakian and J.C. Zorn, Phys. Rev. A, 6 (1972) 2180.
175 J. Berkowitz and J.H.D. Eland, Abstracts X I.C.P.E.A.C. 1977 Commissariat à l'Énergie Atomique, Paris, Vol. 1, p. 110.
176 J.H.D. Eland, Int. J. Mass Spectrom. Ion Phys., 9 (1972) 397.
177 A. Crowe and J.W. McConkey, Phys. Rev. Lett., 31 (1973) 192.
178 R. Stockbauer, Int. J. Mass Spectrom. Ion Phys., 25 (1977) 89.
179 B.A. Huber, T.M. Miller, P.C. Cosby, H.D. Zeman, R.L. Leon, J.T. Moseley and J.R. Peterson, Rev. Sci. Instrum., 48 (1977) 1306.
180 J.A.R. Samson and J.L. Gardner, Phys. Rev. Lett., 33 (1974) 671.
181 M.Ya Amusia, N.A. Cherepkov and L.V. Chernysheva, Phys. Lett. A, 40 (1972) 15.
182 P.G. Burke and K.T. Taylor, J. Phys. B, 8 (1975) 2620.
183 M.Ya Amusia, V.K. Ivanov, N.A. Cherepkov and L.V. Chernysheva, Phys. Lett. A, 40 (1972) 361.
184 W. von Niessen, L.S. Cederbaum and W.P. Kraemer, J. Chem. Phys., 65 (1976) 1378.
185 R.L. Kelly, Atomic Emission Lines below 200 Ångstroms, N.R.L. Rep. 6648, Naval Research Laboratory, Washington D.C., 1968.
186 W.R.S. Garton, M.S.W. Webb and P.C. Wildy, J. Sci. Instrum., 34 (1957) 496.
187 A.W. Potts and T.A. Williams, J. Chem. Soc. Faraday Trans. II, 72 (1976) 1892.
188 J.L. Gardner and J.A.R. Samson, J. Chem. Phys., 62 (1975) 4460.

Chapter 2

Auger electron spectroscopy

S. HOFMANN

1. Introduction and historical development

Auger electron spectroscopy has received rapidly growing attention as a surface analysis tool during the last decade. Its importance
in this field has been so overwhelming that it has become almost
a standard analytical method for the elemental characterization of
solid surfaces and thin films. The main reasons are

(a) the ease of production of Auger electrons combined with
powerful electron analysis devices,

(b) the good overall sensitivity with not too large differences
between the elements,

(c) the absence of first-order chemical effects in most cases,

(d) the ease of combination with noble gas ion sputtering which
allows the determination of well-resolved concentration—depth
profiles.

Numerous review articles on AES have already been published.
The following text on Auger electron spectroscopy from the
analytical standpoint will give special attention to the needs of the
practical analyst. Therefore, theoretical aspects of AES are only
briefly considered, and special emphasis is placed on laboratory
practice, experimental procedures and data evaluation.

Auger electron spectroscopy bears the name of Pierre V. Auger,
who discovered this special type of electron in 1923 [1—3]. Studying
the electron emission induced by X-ray excitation in a Wilson cloud
chamber, he observed electrons emitted with constant energy
independent of the X-ray energy. Further experiments and explana-
tions were summarized in 1926 in his doctoral thesis. As he wrote
in 1975 [4], he was by no means aware of the special meaning of his

findings. For a long time afterwards, the detection of Auger electrons was a special subject in atomic physics and was looked upon as an unpleasant effect in X-ray spectroscopy.

It was only in 1953 that Lander pointed out the possibility of surface analysis by characteristic Auger electron peaks, which he observed in the secondary electron emission spectrum of electron-irradiated solids [5]. But the Auger peaks were too small to be used for analytical purposes. The pioneering work of Harris [6,7] with a 127° energy analyzer in 1965 showed that the detection of the electronically differentiated energy distribution greatly enhances the signal-to-noise ratio of the Auger electron peaks. In 1967, Weber and Peria [8] recognized that the same techniques could be adopted for conventional LEED optics which were already in use in about a hundred laboratories working on surface research. This was the breakthrough of AES as a surface analytical tool and the onset of the rapid development still going on today.

The next major step forward was the introduction of the cylindrical mirror analyzer by Palmberg et al. in 1969 [9]. This device, now generally used in analytical Auger electron spectroscopy, allows an oscilloscope scan of the Auger spectra and the detection of elemental surface concentration down to 10^{-3} of a monolayer. Since that time, a still growing number of commercial instruments of this type has become available.

Further development concentrated on focussing the primary electron beam to improve the lateral resolution and, by scanning the electron beam, to produce Auger electron images of the surface. In 1970, MacDonald [10] showed the principal possibility of high spatial resolution AES, and in 1973, the first scanning Auger microprobe with 5 μm lateral resolution was developed. At present, instruments with submicron spatial resolution are available (see Sect. 12.A).

Parallel with this development, the combination of AES with sputtering techniques revealed its capability for in-depth analysis, first applied by Palmberg and Marcus in 1969 [11]. In the past ten years, there has been a tremendously increasing literature on Auger electron spectroscopy. The work until 1975 has been reviewed by Hawkins [12]. A number of review articles are given in the refs. 13—18.

2. List of abbreviations and symbols

Abbreviations

AES	Auger electron spectroscopy
ESCA	Electron spectroscopy for chemical analysis (X-ray photoelectron spectroscopy, XPS)
GDMS	Glow discharge mass spectroscopy
GDOS	Glow discharge optical spectroscopy
FIM-AP	Field ion microscopy-atom probe
EMP	Electron microprobe
EID	Electron induced desorption
LEED	Low energy electron diffraction
SAM	Scanning Auger microscopy
SCANIIR	Surface chemical analysis by neutrals and ion impact radiation
SNMS	Secondary neutrals mass spectroscopy
SIMS	Secondary ion mass spectroscopy
IMMA	Ion microprobe mass analysis
ISS	Ion scattering spectroscopy
CMA	Cylindrical mirror analyzer
RFA	Retarding field analyzer
SXAPS	Soft X-ray appearance potential spectroscopy
DAPS	Disappearance potential spectroscopy
FWHM	Full width at half maximum (of a Gaussian distribution curve)
XFA	X-ray fluorescence analysis

Symbols

a	parameter defining the layer thickness eroded with constant sputtering rate in the sequential layer sputtering model [nm]
c	atom concentration [cm^{-3}]
c_i	atom concentration of element i [cm^{-3}]
d	thickness of a thin homogenous sandwich layer [nm]
e	electron charge
$g(z)$	resolution function in depth profiling
j_p	primary ion current density [A/cm^2]
k	amplitude of the modulation energy used for derivative signal detection [eV]

l	layer number parameter in the sequential layer sputtering model with escape depth inclusion
n	layer number parameter in the sequential layer sputtering model without regarding escape depth
r	back-scattering factor for electron excitation ($\geqslant 1$)
t	sputtering time [s]
z	depth beneath the original surface [nm]
Δz	depth resolution in sputter profiling [nm]
$\dot{z}$	sputter erosion rate, dz/dt
A	area analyzed [cm^2]
B	bandwidth of the spectrometer [s^{-1}]
E	energy [eV]
E_{kin}	kinetic energy
E_F	Fermi energy
E_W	ionization energy of level W
E_{WXY}	Auger electron energy involving electron niveaus denoted by W, X, Y
E_p	primary electron energy
ΔE	half-width of the energy window of the analyzer
I	intensity of Auger signal (Auger current) in arbitrary units
I_N	intensity of noise signal
I_i	intensity of an Auger signal of element i
I_0	Auger intensity normalization factor
I_s	maximum Auger intensity for a single sandwich sputtering profile
N	total number of layers contributing to surface coverage in the sequential layer sputtering model
N_0	number of atoms per unit area [cm^{-2}]
$N(E)$	number of secondary electrons at energy E: energy distribution of secondary electrons in arbitrary units
$R = \Delta E/E$	energy resolution of the analyzer
S/N	signal-to-noise ratio, noise figure
$S_i/S_j = S_{i,j}$	relative elemental sensitivity of an Auger transition of element i with respect to that of an element j in terms of peak-to-peak amplitude of the $dN(E)/dE$ spectrum
T	total transmission of the analyzer
$U = E_p/E_w$	reduced energy of the primary electrons

$V_{1, ..., 6}$ V_p, V_E, V_c, V_D	} valves of the UHV system
V_M	modulation peak-to-peak voltage of the CMA
w	probability of X-ray fluorescence
X_i	atomic fraction of an element i in a multi-element sample
α_i	Auger yield factor $[cm^2]$
γ	sputtering yield in removed atoms per incident primary ion
σ	standard deviation of a Gaussian integral or error function
σ_W	ionization cross-section for electrons in the W(K, L, M ...)-shell $[cm^2]$
$\rho_{(WXY)}$	probability factor for a WXY Auger transition ($=1-w$)
λ	escape depth of Auger electrons [Å, nm], to be multiplied by the cosine of the take-off angle if used in equations ($\lambda \times 0.74$ for CMA)
ω	frequency of the modulation voltage of the analyzer (fundamental) $[s^{-1}]$
η_i	density correction factor of a sample with respect to a standard
τ	sputtering time to remove one monolayer equivalent [s]
τ_A	time constant of the detection device [s]
φ	take-off angle of the analyzer ($\varphi = 42.3°$ for CMA)
Φ_A	work function of the analyzer [eV]
Φ_s	work function of the sample [eV]
Γ	analyzer transmission factor
Ω	spherical angle [sterad]
θ	coverage of a layer in fractions of a monolayer

3. Brief description of the method

The main parts of an Auger electron spectrometer, shown schematically in Fig. 1, are the electron gun, the target (on an adjustable target holder) and the electron energy analyzer, which are mounted in a vacuum chamber with obtainable vacua levels <100 nPa ($<10^{-9}$ Torr) (ultrahigh vacuum, UHV).

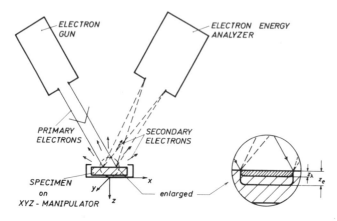

Fig. 1. Principal parts of an Auger electron spectrometer. The enlarged sample cross-section shows schematically the excitation depth of the primary electrons z_e and the (smaller) information depth z_λ determined by the Auger electron energy.

When primary electrons with sufficiently high energy (1—10 keV) from the electron gun strike the target atoms (resulting in ionization of different electron shells) within the excitation depth z_e (100 nm— 1 μm), electrons from other shells can fill up the ionized states, and the energy released by this process is either in the form of a photon (X-ray fluorescence) or is transmitted to another electron in an outer shell (Auger electron) which is emitted from the atom. Auger electrons with definite energies can only be detected if they reach the surface without inelastic scattering in the solid. Their inelastic mean free path, conventionally denoted as escape depth λ, is mainly dependent on the Auger electron energy and lies typically between 0.4 and 4 nm [19—22]. It is also dependent on the respective matrix as seen in the compilation of data given in Fig. 2 [12]. Approximately, above 100 eV λ can be estimated by [229,249]

$$\lambda = 0.2\sqrt{E} \qquad (1)$$

with the electron energy E in eV and λ in monoatomic layers. The surface specificity of AES is a direct result of the small value of the Auger electron escape depth.

The electron energy analyzer can be any device (see Sect. 5.B) which allows the sampling of the secondary electrons and the detec-

94

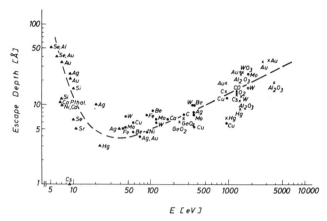

Fig. 2. Experimentally determined mean electron escape depth λ versus electron energy for different materials. Points indicate XPS (×), UPS (●) and AES (▲) determinations: from Brundle [230].

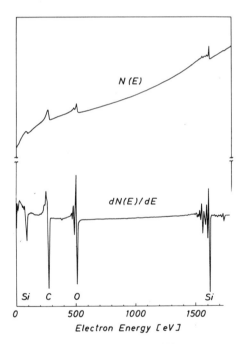

Fig. 3. Energy distribution $N(E)$ and first derivative $dN(E)/dE$ (CMA measurement, E_p = 5 keV, silicon oxide layer with carbon contamination).

tion of their energy distribution function $N(E)$ (Fig. 3). The Auger electrons are superimposed on a large and smoothly varying background consisting of back-scattered primary electrons and inelastically scattered Auger electrons. Therefore, as depicted in Fig. 3, Auger peaks are usually detected by electronic differentiation of the $N(E)$ curve giving $dN(E)/dE$ [17,23] to improve the signal-to-noise ratio.

The detected energy of the obtained Auger peaks is characteristic for the element analyzed (see Sects. 4.B and 7). The area under the $N(E)$ function after background removal is approximately proportional to the number of atoms within the primary beam diameter and the information depth (see Sect. 8, quantitative analysis). This holds also for the $dN(E)/dE$ function if the Auger peak shape is constant with varying Auger intensity.

The rather small escape depth generally restricts the chemical information to the first few atomic layers. Therefore, AES is not adapted to bulk analysis. Even if contamination layers are carefully removed, e.g. by argon ion sputtering (see Sect. 6), the composition of the sample may deviate from the bulk composition due to segregation or preferential sputtering effects (see Sects. 9.B and 10.D).

Since the surface of a sample is steadily reacting with the ambient gas atmosphere, constancy of the surface composition with time may hardly be achieved. To reduce this effect, AES is generally performed in a vacuum chamber at residual gas pressure of reactive gases below 10^{-6} Pa (10^{-9} Torr) (see Sect. 5). A gas admission facility for Ar and an ion gun are necessary for surface cleaning (see Sect. 6.A) and for in-depth profiling (see Sect. 9.B). Furthermore, a residual gas analyzer is useful for deciding whether a detected component is due to adsorption.

4. Theoretical background

There exist some excellent books and papers on the detailed theoretical treatment of Auger transitions [24—26,30], which surpass the scope of this chapter. Correct interpretation of Auger spectra requires at least a basic knowledge of the theoretical background, which will be briefly outlined.

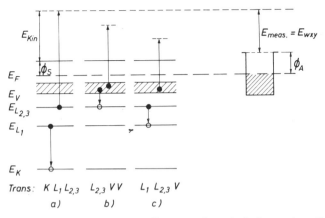

Fig. 4. Schematic energy diagram of typical Auger transitions. (a) Inner shell transition; (b) valence band transition; (c) Coster—Kronig transition. See text for details.

(A) AUGER TRANSITIONS AND AUGER ENERGIES

If an electron from an inner shell is removed by any kind of excitation (X-rays, electrons, ions), the first steps in the reestablishment of energetic equilibrium occur either through photoemission (X-ray fluorescence) or through the Auger effect. Special cases are illustrated in Fig. 4. From part (a) of this figure, the kinetic energy, E_{kin}, of the emitted Auger electron can be derived as follows: the energy gain $E_K - E_{L_1}$ is transmitted to an electron in the $L_{2,3}$ level, which must exceed the energy $E_{L_{2,3}} + \Phi_s$ to be emitted into the vacuum. (Φ_s is the work function of the sample.) At least $E_{L_{2,3}}$ must be corrected due to the extra positive charge of the ionized atom. Therefore, an apparently higher atomic number $Z + \Delta$ applies for the $L_{2,3}$ level [16,25,26] with $\Delta \leqslant 1$. The kinetic energy is

$$E_{kin} = E_K(Z) - E_{L_1}(Z) - E_{L_{2,3}}(Z + \Delta) - \Phi_s \qquad (2)$$

The process is called the $KL_1L_{2,3}$ Auger transition. For generalization, the notation WXY is frequently used to describe a certain Auger transition [16,17,20]. Since the initial and final states are the same for WXY and WYX transitions, a similar charge correction for the X and Y levels will better account for their identical energy [27]. The energy measured is E_{kin} reduced by the difference in the work functions of the analyzer and the sample, $\Phi_A - \Phi_s$ (Fig. 4).

$$E_{WXY} = E_W(Z) - E_X(Z') - E_Y(Z') - \Phi_A \tag{3}$$

To estimate the effective charge Z', an empirical approach has been found useful [27]: an average energy between the value of the atomic numbers Z and $Z + 1$ is taken for the E_X, E_Y binding energies

$$E_X(Z') = \tfrac{1}{2}[E_X(Z) + E_X(Z + 1)]$$

and

$$E_Y(Z') = \tfrac{1}{2}[E_Y(Z) + E_Y(Z + 1)]$$

Using this approximation, we get for the Auger electron energies [27]

$$E_{WXY}(Z) = E_W(Z) - \tfrac{1}{2}[E_X(Z) + E_X(Z + 1) + E_Y(Z) + E_Y(Z + 1)] - \Phi_A \tag{4}$$

This energy can be determined from X-ray tables of electron energies [28] with a typical accuracy of a few electron volts [26,27].

Since at least two electrons in the L shell are required for the Auger process depicted in Fig. 4(a), Be is expected to be the lightest element to show an Auger transition. However, if metallic lithium is considered, the L shell will broaden to a valence band and an Auger transition as in Fig. 4(b) is possible. In their normal states, H and He give no Auger signal.

The width of the Auger lines is limited by the transition time, which is typically of the order of 10^{-15} s. Due to the uncertainty principle, ΔE_{WXY} is of the order of a few eV. In the case shown in Fig. 4(b), a valence band transition of the type WVV will lead to an enlarged line width of about twice the valence band energy width. The energy distribution reflects the band structure. Valence band transitions are important in regarding chemical effects (see Sect. 7.C).

Another type of transition, the Coster—Kronig transition WWX [29,30], gives a significant line broadening due to its small transition time ($\leqslant 10^{-16}$ s).

The possible Auger electron energies may be calculated from eqn. (2) from all the energetically possible transitions using tables of electron energies [28], because there are no strong selection rules as there are for X-ray transitions. In general, theoretical calculations are in fairly good agreement with the values from measured Auger transitions [16,31]. [Measured Auger electron energies are generally referred to the negative peak location in the derivative spectrum [32] (see Sect. 5.B) which must be considered here.] A diagram of the principal Auger transitions observed from the elements is shown in Fig. 5 [23,33]. A very useful catalog of calculated Auger transition based on eqn. (4) has been given by Coghlan and Clausing [34]. An example for sulphur is shown in Table 1 [34].

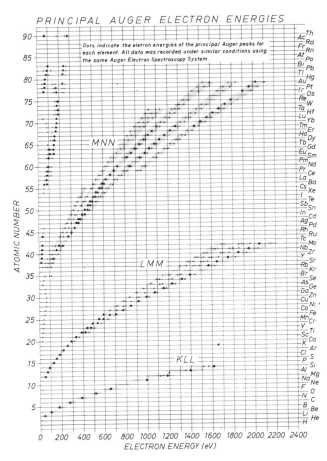

Fig. 5. Principal Auger electron energies of the elements. From Joshi et al. [17] by courtesy of Elsevier Scientific Publishing Company.

TABLE 1

Calculated Auger energies for sulphur and term multiplicities using eqn. (4) ($\Phi_A = 0$) from Coghlan and Clausing [34] Levels used for S ($Z = 16$)

Orbital	Population	Z-energy [eV]	($Z + 1$) energy [eV]
K	2	2472.00	2823.00
L_1	2	229.00	270.00
L_2	2	165.00	202.00
L_3	4	164.00	200.00
M_1	2	16.00	18.00
M_{23}	4	8.00	7.00

Vacancy level	Interaction levels		Auger energy [eV]	Norm- multiplicity
L_1	L_2	M_1	28.50	12
L_1	L_3	M_1	30.00	25
L_1	L_2	M_{23}	38.00	25
L_1	L_3	M_{23}	39.50	50
L_3	M_1	M	130.00	25
L_2	M_1	M_1	131.00	12
L_3	M_1	M_{23}	139.50	50
L_2	M_1	M_{23}	140.50	25
L_3	M_{23}	M_{23}	149.00	100
L_2	M_{23}	M_{23}	150.00	50
L_1	M1	M_1	195.00	12
L_1	M_1	M_{23}	204.50	25
L_1	M_{23}	M_{23}	214.00	50
K	L_1	L_1	1573.00	12
K	L_1	L_2	2039.00	12
K	L_1	L_3	2040.50	25
K	L_2	L_2	2105.00	12
K	L_2	L_3	2106.50	25
K	L_3	L_3	2109.00	50
K	L_1	M_1	2205.50	12
K	L_1	M_{23}	2215.00	25
K	L_2	M_1	2271.50	12
K	L_3	M_1	2273.00	25
K	L_2	M_{23}	2281.00	25
K	L_3	M_{23}	2282.50	50
K	M_1	M_1	2438.00	12
K	M_1	M_{23}	2447.50	25
K	M_{23}	M_{23}	2457.00	50

(B) AUGER INTENSITIES

The intensity of an observed WXY Auger transition for a given primary electron current will depend on the ionization cross-section, σ_W, and the Auger yield per vacancy ρ (WXY). The latter quantity is given by the Auger transition probability $(1 - w)$, where w is the probability of a radiative transition, that is the X-ray fluorescence yield [16,25,26]. Figure 6 [28] shows a plot of the Auger yield versus atomic number for K- and L-shell excitations together with the binding energies. The Auger yield is unity for light elements and decreases with increasing Z at first for the K- and then for the L-shell ionization etc. This is one reason why the more sensitive Auger transitions analytically applied are well below ~ 2.5 keV, that means KLL transitions for elements $Z = 3{-}14$, LMM transitions for elements $Z = 14{-}40$, and MNN transitions for the heavier elements [17] (see Fig. 5).

The ionization cross-section, σ_W, is strongly dependent on the primary beam energy, E_p. A universal function has been derived from various theoretical approximations with the use of relative energy $U = E_p/E_W$ as the decisive parameter as shown in Fig. 7. In the energy range of interest in AES, the Gryzinski calculation [36], has been found to be appropriate [37]. $U = 1$ denotes the ionization threshold.

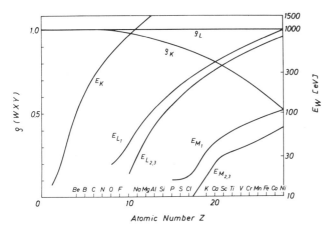

Fig. 6. Ionization energies E_K, E_{L1}, $E_{L2,3}$, E_{M1}, $E_{M2.3}$ (right-hand scale) and Auger transition probabilities ρ_K, ρ_L as a function of the atomic number. From Bauer [35] by courtesy of Pergamon Press.

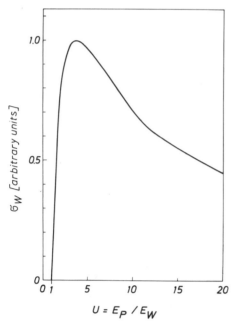

Fig. 7. Ionization cross-section σ_W for inner shell ionization as a function of incident electron energy $U = E_p/E_W$; after Gryzinski [36].

A maximum for σ_W is obtained at $U = 3$—5. For higher energies, the ionization cross-section decreases gradually. The fact that the absolute values of σ_W strongly decrease with the binding energy [38,39] again suggests that, for the heavier atoms, L- and M-shell Auger spectra should be used. Experimentally, the primary energy dependence of Auger intensities is complicated by the contribution of back-scattered primary electrons [32,37,38,40] as discussed in Sect. 8.

An exact calculation of absolute intensities is complicated by the coupling schemes governing the transition from a single-ionized to a double-ionized state of the atom [16,17,26,30]. For a first-order approximation, the calculation of relative intensities of the Auger transitions of each element was performed by Coghlan and Clausing (Table 1) [34]. They used the products of the number of electrons in the three levels involved in the transition, normalized to a maximum value of 100 for the largest product of a given element. This number, the "normalized multiplicity", gives a crude but helpful means to identify the most prominent Auger transitions expected in qualitative analysis (see Sect 7.A).

5. Instrumentation

A basic requirement in instrumentation is a suitable vacuum chamber in which a sufficiently low residual gas pressure can be achieved. Although 10^{-2} Pa (10^{-4} Torr) is sufficient for a mean free path of the Auger electrons of the order of the analyzer dimensions, vacuum levels of $<10^{-7}$ Pa ($<10^{-9}$ Torr) should be maintained. The main reason for this restriction is the influence of contamination layers on the Auger yield of the sample due to the high surface sensitivity of AES. Contamination layers build up quite readily by the adhesion of reactive residual gas components (e.g. CO, H_2O) to the sample surface. If the adhesion coefficient is ~ 1, a rough estimate gives about one atomic monolayer per second at 10^{-4} Pa (10^{-6} Torr). Therefore, the pressure should be well below this figure to ensure a complete measurement cycle before reactions with residual gases alter the surface. To protect the sample surface from carbon contamination due to cracking of hydrocarbons by the electron beam [41] an oil-free vacuum system is necessary. Therefore, ion pumps or turbomolecular pumps are most frequently used.

To control the gas composition, a quadrupole mass analyzer is desirable. An ion gun with suitable gas inlet device for surface cleaning and depth profiling (see Sect. 9) is generally attached.

(A) EXCITATION METHODS

The prerequisite for an Auger transition to occur is an atom ionized in an inner shell. Therefore, any means by which such an ionization can be obtained may be used for the excitation of Auger electron emission. Historically, X-ray excitation was the first method used [4,42]. Because of its limited excitation density, it is more commonly used in photoelectron spectroscopy [22,43] than in AES. However, Auger electrons are readily detected in ESCA [43,205,206]. Heavy particle bombardment has also been used to excite Auger electrons [44—46]. Again, the obtainable current densities, in spite of favourable ionization cross-sections, prevented a general application. Recently, the primary ion current density of commercially available ion guns with ion energies of $\geqslant 5$ keV was increased up to $>100\,\mu A/cm^2$, so that easily detectable ion-induced Auger spectra are generated [47]. They show differences with X-ray or electron-induced spectra and are expected to give valuable additional information, especially with respect to chemical bonding [47,48] in the near future.

The most commonly used excitation technique is electron bombardment. Electron guns have the advantages of easy construction and maintenance, high beam intensity and capability of focussing and $x-y$ deflection. These latter points are the main objectives of the recent development towards scanning Auger microscopy [49] with high spatial resolution down to 30 nm [50].

The role of primary electron energy and current density in AES is discussed in detail in Sects. 7, 8 and 10.

Other sources of Auger electrons, like electron capture and internal conversion in a radioactive element, are less important for analytical purposes [25].

(B) ELECTRON ENERGY ANALYSIS

The various types of instrumentation comprise two main branches: magnetic and electrostatic analyzers. Due to their complexity in construction, magnetic analyzers are only used if very high energy resolution ($\leqslant 0.01\%$) [25] is requested. They are not used in normal Auger spectrometers and therefore will not be discussed here (a thorough discussion may be found in ref. 25).

Electrostatic analyzers comprise two main subdivisions: (a) retarding field analyzers and (b) electrostatic deflection analyzers. The functional qualities or merits of the various types of spectrometer may be given in terms of

(1) total transmission, T, that is the fraction of emitted monoenergetic electrons which pass the detector slit;

(2) acceptance angle, Ω, (point source) and acceptance area, A, (extended source) of the source

$$TA = L$$

where L is the overall luminosity [25];

(3) energy resolution in terms of $R = \Delta E/E$, where ΔE is the full width at half maximum intensity (FWHM) of the monoenergetic electron line and E the energy value of this line.

(4) the energy filtering quality (or electron-optical quality) which

is given by the ratio of its transmission to its resolving power, T/R. This figure mainly determines the maximum obtainable signal-to-noise ratio, S/N (sensitivity). The limiting factor of the S/N ratio is the shot noise current, I_N, given by the collected current I and the bandwidth B according to [13,51,52]

$$I_N = (2eIB)^{1/2} \qquad (5)$$

$B \propto 1/\tau_A$, where τ_A denotes the time constant for the spectrometer. Shot noise is due to the discrete quantification of the electric charge. Therefore, counting statistics ($\propto I^{1/2}$) pose an ultimate error bar, which cannot be avoided by any sophistication in measurement devices.

Only a brief outline of the fundamental principles of the electrostatic detection methods, subdivided into retarding field analyzers (RFA) and electrostatic deflection analyzers will be given here. A number of detailed discussions are given in the literature [16,25,53—57].

The most simple method (RFA) which was first successfully applied in the work of Franck and Hertz on the detection of ionization potentials in 1925, is that of placing a retarding potential on a grid before the detector. In this way, Lander [5] used a three-grid hemispherical system customary in low energy electron diffraction (LEED) studies for analyzing Auger electron spectra from electron

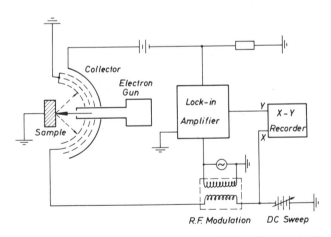

Fig. 8. Schematic arrangement for AES with a retarding potential analyzer (LEED—Auger system).

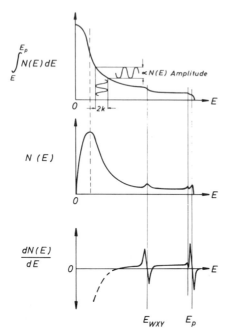

Fig. 9. Principle of the different modes of operation for the retarding field analyzer (RFA); after Taylor [13].

bombarded surfaces. Weber and Peria [8], Harris [6] and Taylor [51,57] developed the standard technique to get the energy distribution curve $N(E)$ and the differentiated $N(E)$ curve, $dN(E)/dE$, by electronic differentiation. Figure 8 shows the block diagram of the LEED Auger device [58].

The grounded grids are necessary to ensure field-free space between sample and the retarding field grid and to shield it from the collector potential, which is positive to supress secondary electrons. (Two grids at retarding voltage are sometimes used to improve the energy resolution [8].) Increase of the negative potential, V, of the middle grid by the d.c. sweep cuts off the incoming electrons with energies below $E_0 = eV$ (high pass filter [52]). A plot of the collector current versus retarding field, E, yields the integral of the kinetic energy distribution $N(E)$ as depicted in Fig. 9 [13].

$$I(E_0) \propto \int_{E_0}^{E_p} N(E) \, dE \tag{6}$$

This function must be differentiated to obtain the energy distribution $N(E)$, which is done by applying a small a.c. modulation voltage to the retarding grids and detection of the output by a lock-in amplifier (Fig. 8). If the energy is modulated to $E = E_0 + k \sin \omega t$, the collector current can be written in a Taylor series approximation around E_0 [13,15,51]

$$I(E_0 + k \sin \omega t) \propto I(E_0) + N(E_0) \, k \sin \omega t - \frac{1}{4} \frac{dN(E_0)}{dE} \, k^2 \cos 2\omega t \pm \ldots \tag{7}$$

The modulation amplitude, k, should be small enough so that the k^3 and higher order terms can be neglected [15]. Equation (7) is illustrated in Fig. 9. It shows that the energy distribution, $N(E)$, is proportional to the amplitude of the fundamental modulation frequency term at ω, and the first derivative $dN(E)/dE$ is obtained by detection at the second harmonic frequency 2ω, $I_{2\omega}(E) \propto dN(E)/dE$. Experimental values of ω are usually of the order of 10^4 Hz.

Obviously, optimum resolution is obtained for a very small modulation voltage. On the other hand, reasonable output signals require a sufficiently high k, which may lead to severe deviations from proportionality due to the higher order terms in eqn. (7). In AES work, a useful compromise, found empirically for optimum modulation voltage lies generally between 1 and 10 eV, depending on the Auger peak width [51]. Figure 9 shows the advantage of the reduced background slope if the second harmonic signal amplitude is detected.

Since there is still a varying slope in the $N(E)$ background, the double peak shape of the differentiated Auger line is not symmetrical and it is very difficult to locate the exact position of the Auger electron energy where $dN(E)/dE$ goes through the background level. Therefore, it is customary to take the Auger energy at the minimum of the differentiated spectrum [15].

Electrostatic deflection analyzers differ from the RFA by their electron optical properties, that means an image of the source is established at the entrance slit of the detector. There are four types of deflection analyzers [25], the parallel plate analyzer [59], the $127°$ sector analyzer [60] (both single focussing devices), the $180°$ spherical analyzer [61] and the cylindrical mirror analyzer (CMA) [9,54,55]. Only the latter two, which are double focussing devices [25], have practical importance in analytical electron spectroscopy. Whereas the spherical analyzer is commonly used in ESCA devices

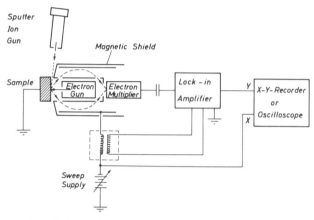

Fig. 10. Schematic arrangement for AES with a cylindrical mirror analyzer (CMA).

[32,206], the CMA has become the most prominent Auger analyzer.

The general features of an Auger electron spectrometer with CMA are depicted in Fig. 10. The CMA consists of two concentric cylinders with two gridded radial apertures which allow only those electrons to be detected (by an electron multiplier) which can follow the path shown by the broken lines in Fig. 10. For a certain potential, V_a, of the outer cylinder, only electrons within a small energy window ΔE around the pass energy $E = eV_e$ will be detected. (The ratio of V_e/V_a is determined by the geometry of the CMA [9,25].) Thus, in contrast to the RFA, which is a high pass filter, the CMA is a window device, or a band pass filter. The current at the output $I(E)$ is given by [26]

$$I(E) \propto \Delta E N(E) = R E N(E) \tag{8}$$

where ΔE is the width of energy window and $R = \Delta E/E$ the resolution of the CMA.

The differentiated function $dN(E)/dE$ is again obtained by applying a modulation voltage $V_M \sin \omega t$ to the outer cylinder (Fig. 10), but now detecting the first harmonic, ω, of the output current. Differentiation of eqn. (8) gives, for the amplitude of the derivative signal

$$I_\omega(E) \propto R k [N(E) + E \, dN(E)/dE] \tag{9}$$

The additional $N(E)$ term in eqn. (9) and the electron multiplier gain

108

function are disadvantageous for quantitative work with respect to the RFA, especially at low energies. However, a comparison of the operation principles of both devices shows the unique advantages of the CMA:

(a) Detection limit. As pointed out above, the noise figure puts an ultimate limit on sensitivity. Since only electrons in the window ΔE are collected by the CMA, the currents are typically 10^{-4} times smaller than in the RFA device giving a noise smaller by a factor of 100 according to eqn. (5) [51]. Small primary electron currents are found very useful if electron beam induced alterations of the surface occur (see Sect. 10.C). Equation (5) also shows that the sampling time τ_A may be reduced by a factor of 10^{-4} with respect to RFA at the same detection level. This enormous increase in detection speed allows oscilloscope inspection of an Auger spectrum using CMA. Due to its focussing property, a small spot is formed at the entrance slit of the detector so that an electron multiplier can be used for sensitive detection.

(b) Transmission. A high transmission is the prerequisite of high detection sensitivity (see Sect. 8.C). In a hemispherical grid device, transmission would be expected to be almost 100%. However, due to the non-zero grid currents, $T \sim 10\%$ is a typical value [57]. About the same transmission is achieved with the CMA [15]. The luminosity, $L = T\Omega$, due to the large acceptance angle, is higher for the RFA.

(c) Resolution. The resolution of the hemispherical grid RFA is given by the geometry of the grids and by the applied modulation voltage V_M [57]. As for the CMA, $\Delta E/E$ is an apparatus constant for fixed V_M. The CMA resolution is primarily dependent on the acceptance angle with respect to the analyzer dimensions. A typical figure is $R = \Delta E/E = 0.5\%$ for most commercial systems.

The T/R value appears to be not very different for both analyzer devices for medium Auger energies. However, the outstanding signal-to-noise figure of the CMA has made this device the most feasible instrument for analytical AES.

6. Practice of AES

In this section, the general run of an AES analysis with a commercial CMA instrument is outlined, with emphasis on practical hints for the operator.

A natural limit of specimen size and shape is given by the dimensions of the vacuum chamber. The analysis area is determined by the primary electron beam diameter (generally $\leqslant 100\,\mu m$) within the acceptance region of the analyzer. To this focus point of the analyzer, the interesting part of the sample to be analyzed has to be adjusted. This poses difficulties for irregular shaped samples; flat samples with smooth surfaces are desirable. A certain roughness can be tolerated for qualitative analysis but is unfavourable for quantitative work (see Sects. 8 and 9).

(A) SPECIMEN PREPARATION

The preparation of the specimen depends on the purpose of the AES analysis. If one is interested in the composition of the uppermost layers of the sample in the as received state, e.g. in contamination or corrosion studies, no cleaning procedure should be applied and the specimen is directly mounted on the sample holder. If deeper layers are of interest, or the sample is used for more fundamental studies (e.g. segregation, diffusion, adsorption, etc.), at least a degreasing procedure is recommended: preliminary rinsing in acetone or freon, then in ethylene (preferably in an ultrasound vibration tub) and subsequently in doubly distilled water. Drying in an incubator completes the cleaning procedure. The specimen is now mounted on a generally used carousel-type target holder together with other samples or standards. For this procedure, a laminar flow box is preferred to avoid dust contamination. Handling all parts which will be placed inside the UHV chamber should be done using plastic gloves and carefully cleaned forceps, screwdrivers etc.

Even in the case of the cleaning procedures described above, a certain contamination from the air and from residues of the cleaning agents is inevitable. Auger analysis of the sample after introduction into the vacuum chamber shows, in any case, a carbon and oxygen signal, sometimes accompanied by sulphur and chlorine (Fig. 11). To remove these contaminants, additional cleaning techniques in the chamber must be applied. The most customary technique is argon ion bombardment with an auxiliary ion gun, which has become standard equipment in an Auger spectrometer (see Fig. 10). Since sputter removal and Auger spectrometry can be done simultaneously at the same spot, this technique gives immediate information on the in-depth distribution of composition (see Sect. 9.B).

110

(B) PREPARATION OF STANDARDS

Qualitative as well as a semiquantitative analysis can be performed using standard spectra and relative sensitivity factors of the elements, which are compiled in handbook form [27] (see Sect. 8.B). A more exact approach is to rely on the relative elemental sensitivities determined with pure substances under the experimental conditions of one's own instrument. For solid metals, for instance, this is no problem. The cleaning procedure as outlined in Sect. 6.A has to be applied (see Fig. 11). With elements not available in the solid state at room temperature (e.g. mercury, oxygen etc.) one must rely on stable alloys or compounds of otherwise known composition. However, in this case, the surface composition can deviate appreciably

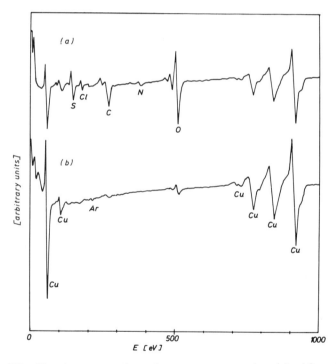

Fig. 11. Auger spectra of a copper sample. (a) After transfer from the atmosphere into the vacuum chamber; (b) after sputter cleaning by 1 keV Ar^+ ion bombardment (eroded depth ~30 nm).

References pp. 166—172   111

from the bulk composition due to the sputter cleaning procedure (see Sect. 10.D).

The influence of different methods to obtain a clean surface (cleaving, sputtering and scribing) with known bulk composition on the Auger yield was demonstrated by Braun and Färber [62—64]. Figure 12 [63] shows an example for Ag—Cu alloys. The normalized Auger peak-to-peak heights [$Ag_n = Ag_{meas.}/(Ag_{meas.} + Cu_{meas.})$] are plotted against bulk composition for the different treatments. The differences for Cu 60 eV and Cu 920 eV are mainly due to the different escape depths (about a factor of 4 higher for Cu 920 eV than for Cu 60 eV [82]; cf. Fig. 26). The scribing technique [63] is thought to give no deviation of the surface composition, since no material parameters like tensile strength or selective sputtering of the elements will influence this cleaning technique. Even in this case, a plot of the measured Auger peak-to-peak height ratios versus the atomic compositions as shown in Fig. 13 [63] shows a deviation from linearity, which will be due to a nonlinear dependence of the escape depth or the back-scattering factor. The influence of these

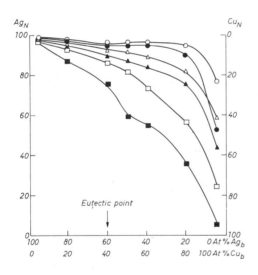

Fig. 12. Normalized Auger peak-to-peak heights (Ag_n, Cu_n) versus bulk concentration of Au—Ag alloys for different treatments. ○, Cleaved, Cu 920 eV; ●, cleaved, Cu 60 eV; △, scribed, Cu 920 eV; ▲, scribed, Cu 60 eV; □, sputtered, Cu 920 eV; ■, sputtered, Cu 60 eV. From Braun and Färber [63] by courtesy of Elsevier Scientific Publishing Company.

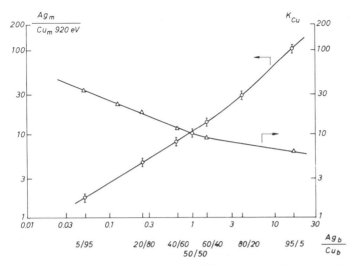

Fig. 13. Measured APPH ratio Ag_m/Cu_m versus bulk composition ratio Ag_b/Cu_b and the relative calibration factor for Cu $K_{Cu} = (Ag_m/Cu_m)/(Ag_b/Cu_b)$ for scribed samples, after Braun and Färber [63] by courtesy of Elsevier Scientific Publishing Company.

parameters on quantitation is discussed in Sects. 8 and 9.

Other methods of standard sample preparation frequently applied comprise evaporation or adsorption of thin elemental layers in situ [65,66] or sputter deposition of alloys with known composition [67,68,219].

(C) SET-UP OF THE SPECTROMETER

We refer to a typical commercial instrument (Physical Electronics, Inc.) with a cylindrical mirror analyzer (Fig. 8).

The vacuum system is shown schematically in Fig. 14. The UHV chamber, which contains the sample holder, the CMA and the ion gun, is connected to a 200 l/s ion pump via the poppet valve, V_p. For introduction of the samples, the UHV chamber is brought up to atmospheric pressure by dry high purity nitrogen from a steel bottle with a pressure reduction valve through the valves V_L and V_E. (Dry nitrogen is essential because water vapour is one of the most prominent species in the residual gas if no bake-out is applied.) Meanwhile, the ion pumps are working with the poppet valve closed at a

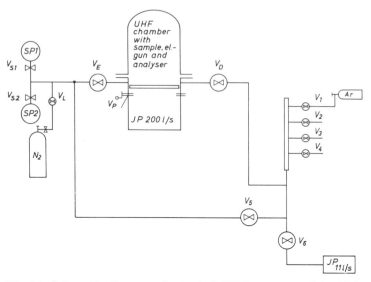

Fig. 14. Schematic diagram of a typical UHV system used in AES. For description, see text.

total pressure of $<10^{-4}$ Pa ($<10^{-6}$ Torr), and the sorption pumps SP1 and SP2 have been chilled for about 20 min with liquid nitrogen and closed valves V_{S1} and V_{S2}. Pump-down starts with the closure of V_L and the opening of V_{S1}. After 2—3 min, 26 Pa (200 mTorr) should be obtained and V_{S1} is closed. In about two minutes, the pressure should not rise if there is no leak in the chamber. The pump-down procedure is continued opening valve V_{S2}, and in a few minutes the chamber pressure will be below 0.8 Pa (5 mTorr). After closing valve V_{S2}, the poppet valve V_p is slowly opened so that the pressure indicated by the current of the ion pumps will not exceed 1 mPa (10^{-5} Torr). Eventually, the poppet valve is fully opened and the ion pumps (assisted by a titanium getter pump) will evacuate the specimen chamber down to the 100 nPa (10^{-9} Torr) region in about 1—2 h. With non-reactive samples (e.g. noble metals, stainless steel, ceramics) AES investigations can be performed in this pressure range with reasonable accuracy. For more surface active species (e.g. Group IVA, VA and VIA metals) and careful quantitative analysis a base pressure in the 10 nPa (10^{-10} Torr) region and below is highly desirable. Depending on the overall state of the vacuum system and the cleanliness of the specimens, this may be obtained even without

114

baking by pumping overnight. Baking (normally to 250°C) with an external heater shroud is highly recommended if the thermal stability of the specimens and the purpose of the investigation allow the thermal treatment.

In Fig. 14, a gas handling system is depicted. It can be evacuated in a similar manner to the UHV chamber by sorption pumping through valve V_5 and by a small ion pump through V_6. To provide the argon needed for sputter cleaning, high purity Ar is introduced to the manifold through V_1 (at closed V_5 and V_6) and then to the working chamber through the needle valve V_D. The valves V_2, V_3, V_4 can be used for other gases, e.g. to perform adsorption studies.

(D) MEASUREMENT SEQUENCE

Having set up the instrument to the working conditions, the first Auger spectrum from a standard or a specimen can be recorded. With the CMA instrument, a prerequisite is careful adjustment of the sample in front of the analyzer. If an integral coaxial electron gun (without x—y deflection) is available, the electron beam will hit the sample within the acceptance area of the analyzer. A typical distance from sample to analyzer entrance is between 5 and 10 mm. The adjustment of the optimum distance is performed looking at the oscilloscope picture of the elastic peak and optimizing its peak-to-peak height. Care should be taken to have optimum phase adjustment at the lock-in amplifier and good focussing of the electron beam. The adjustment procedure is more complicated when working with a grazing incidence gun or a coaxial gun with x—y deflection plates. Here, the electron beam has to be aligned to the acceptance area of the analyzer.

The parameters of the gun and analyzer should be chosen with respect to the analysis problem. An electron beam current of about $10\,\mu A$ will be adequate for general purposes; higher beam currents are only tolerable for metallic samples (when high sensitivity is required) because of increased Joule heating. Lower currents are recommended especially for samples where beam effects are expected (e.g. in oxides, see Sect. 10.C). Time constant and sweep velocity should be in accord [57].

The electron gun voltage should be set at a standard value of 3 keV (if a spectrum up to 1000 eV is recorded) or 5 keV (if a spectrum up to 2500 eV is recorded) and in special cases (scanning

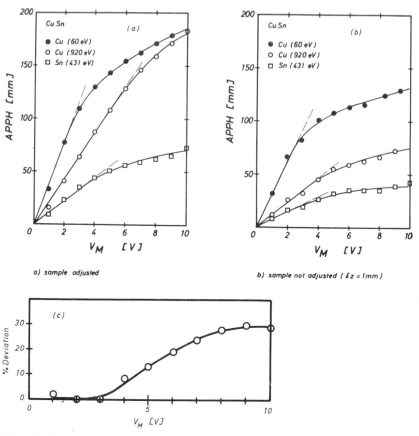

Fig. 15. Measured Auger peak-to-peak heights (APPH) as a function of the modulation voltage, V_M, applied to the CMA for a Cu—Sn alloy. (a) sample well adjusted in front of the analyzer; (b) sample moved 1 mm in the direction of the analyzer (unadjusted); (c) deviation of the APPH ratio for Cu (920 eV)/Cu (60 eV) from (a) versus modulation voltage V_M.

Auger spectrometer with extremely focussed beam) up to 10 keV. For these figures, standard spectra have been recorded [33]. It should be noted that, for special problems, e.g. ultimate sensitivity for low energy Auger peaks (<500 eV), the primary electron energy is preferably reduced. Figure 7 shows the reason for the choice of electron gun voltage. The excitation function goes through a maximum at about $E_W/E_p \sim 3–5$, where E_W is the energy of the ioniza-

116

tion level. It should be noted, however, that back-scattering effects tend to raise the decrease of the ionization function at higher primary energies (see Sect. 8.A).

The peak-to-peak modulation voltage, V_M, should be chosen between 1 and 10 V taking account of the fact that low V_M gives high energy resolution but low sensitivity and vice versa. Therefore, for the first inspection of a spectrum, a high modulation voltage (e.g. 8 V) is generally used (compare Fig. 15).

A comparison of the detected peak heights of an element with a standard sample can only be made if the oxide and/or contamination layer is removed by sputtering with an argon ion gun. For this purpose, the ion pumps are shut off and argon (or another noble gas) is introduced into the sample chamber through a needle valve V_D (Fig. 14) up to a typical pressure of 6.7 mPa (5×10^{-5} Torr). The ion gun is set in operation and sputtering is performed until the oxygen and carbon (and in some cases sulphur, chlorine and nitrogen) peaks normally obtained at any metallic surface have almost vanished. Figure 11 depicts an Auger spectrum of a pure copper surface as introduced (a) and after having sputtered away about 300 Å (b) (compare Sect. 9.B). Only in state (b) may the Cu sample serve as an elemental standard.

(E) ADJUSTMENT AND CALIBRATION PROBLEMS

Using a CMA for Auger analysis, problems in adjustment generally arise because of the focussing properties of this device. If the optimum emission point on the z-axis (symmetry axis) of the CMA is not achieved, transmission (and resolution) will be reduced. This is demonstrated in Fig. 15 for a Cu—Sn alloy. Figure 15(a) shows the dependence of the peak-to-peak heights of the most prominent Auger signals of Sn (451 eV) and Cu (920 eV, 60 eV) on the modulation voltage for optimum adjustment. In Fig. 15(b), the same dependence is given for a deviation of 1 mm on the z-axis from this condition. A reduction of all peaks is recognized, but unfortunately, mainly due to the different peak width, the decrease in yield is different for the respective peaks; the apparent relative sensitivities have changed. Therefore, any reliable calibration presumes an optimum geometrical adjustment.

A further problem in quantification may be the modulation voltage dependence of the different Auger peaks. For small V_M, it is linear as expected (see Sect. 4.C). If V_M approaches the FWHM of an Auger peak, however, overmodulation occurs and the signal height tends to saturate. This point is reached at about $V_M \sim 3 \times FWHM$ [32]. Fig. 15(c) gives an illustration of the deviations in peak-to-peak height ratios which may occur. For quantitative analysis, a small modulation voltage should be chosen. On the other hand, if a certain overmodulation is applied, slight peak shape changes in the course of an experiment (see Sects. 4.D and 7.C) will be less important with respect to relative peak-to-peak heights (see Sect. 8) [250].

A further problem arises from nonlinearities in the energy scale due to false adjustment [70], which should not be interpreted as a real "shift" in energy as sometimes encountered in chemical bonding effects (see Sect. 7.C). The worst case can be encountered if the sample position with respect to the CMA is changed during the course of the experiment, e.g. by mechanical shock or thermal expansion (e.g. heating up a thin sample which may lead to bending).

Extreme care should be taken to avoid any stray magnetic field. Even the earth's field can cause difficulties in the low energy region. If the sample is heated by direct current, the peak heights will be influenced by the magnetic field in a manner difficult to predict. Figure 16 shows an example, where an oxidized niobium ribbon was heated by d.c. current. The measured peak heights had to be corrected by this "calibration" curve to give peak heights suitable for subsequent quantitative analysis [69].

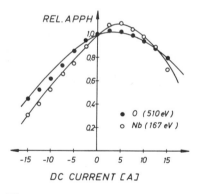

Fig. 16. Magnetic field influence on the measured Auger signal. Relative Auger peak-to-peak heights (APPH) of O and Nb as a function of the d.c. current through an oxidized niobium ribbon.

118

A general check for correct operation of the electron gun and the analyzer are given in the instrument manuals and will not be discussed here. The geometrical adjustment together with the energy scale calibration is done by inspection of the elastic peak with a calibrated electron gun voltage. Optimizing the elastic peak performance also provides a check of optimum phase adjustment of the lock-in amplifier. The ratio of the energy difference of the maximum and minimum of the elastic peak and the peak energy directly give the analyzer resolution. Poor resolution and low signal-to-noise figures for the Auger peaks as well as energy shifts can be due to contamination of the analyzer. This may occur, for example, by evaporation or sputtering procedures performed within the UHV chamber [71]. Low gain, even at high multiplier voltage, may be a hint of old or contaminated multiplier dynodes.

In the case of poor electrical earth connection or insulating specimens, peak energies will shift drastically and will be unstable with time (see Sect. 10.B).

Of course, all effects discussed in Sect. 6.E have to be carefully considered. For a correct set-up of the instrument, Auger peaks above 300 eV should lie within a few eV of the values tabulated [31,33,34].

7. Qualitative analysis

An Auger spectrometer output generally consists of a voltage proportional to $dN(E)/dE$ (y-axis) over the electron energy E (x-axis) as depicted, for example, in Fig. 11. Hence, qualitative analysis is straightforward by comparison of the negative peak energies of the most prominent peaks with tabulated energies from standard spectra of the elements [33,34]. Small deviations may be due to chemical bonding effects (see Sect. 7.C). Sometimes, characteristic loss peaks are observed in the recorded spectrum. They are mainly caused by core level ionization of the primary electrons. Since their energy difference from the primary electron peak is constant, they can be distinguished from Auger peaks by variation of the primary beam voltage. The Auger peaks remain fixed on the energy scale whereas the loss peaks move with the primary beam energy [13,72].

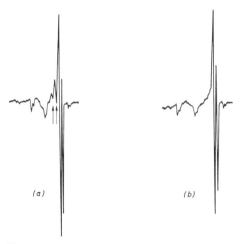

Fig. 17. Auger spectra of Sn in different states. (a) Bulk Sn specimen; (b) Sn segregation layer (about 1 monoatomic layer thickness) at a Cu surface [75].

The excitation of plasmon oscillations gives rise to characteristic loss peaks at the low energy side of an Auger peak [13,61,73,74]. A check of the occurrence of loss peaks can be made by setting the primary electron energy at the value of the observed Auger peak. Since plasma oscillation energies (of the order of 10 eV) are characteristic for the electronic structure, they can give a hint to chemical composition (see Sect. 7.C). In special cases, a distinction between the spectra of a bulk specimen and a thin overlayer is possible by the occurrence of plasmon losses only in the bulk specimen, as shown in Fig. 17 for a bulk Sn sample and for a Sn segregation layer of only one monolayer thickness on a Cu sample [75].

(A) "FINGERPRINT SPECTRA"

As stated in Sect. 4, the Auger energies of the elements may be quite well predicted. Since the relative intensities and peak width may deviate from calculations based on first principles [26], a direct identification of the species in a multielement specimen by using only tabulated energies is rather difficult. Besides the energies, the shape of the Auger spectrum of each element is characteristic and is a reliable guide for qualitative analysis. Standard "fingerprint" spectra of the elements [33] in most cases allow a straightforward identification of the elements present in the sample.

(B) PROBLEMS IN MULTIELEMENT ANALYSIS

The general problem in the analysis of a sample consisting of a large number of elements arises from overlapping of certain Auger peaks in their spectra. Since many of the heavier elements ($Z > 35$) have complicated Auger spectra in the region up to 300 eV, the recognition of lighter elements (C, N, O) possessing only one Auger intense transition in this region can be difficult. An example is given in Fig. 18 for sulphur on a gold surface. The $L_3M_{2,2}M_{2,3}$-sulphur peak at 150 eV overlaps the gold peak at almost the same energy (Au $N_5N_7N_7$ transition [26]). A decision whether sulphur is present can only be made by comparison with the Auger spectrum of pure gold [Fig. 18(b)], which also allows an estimate of the quantity of sulphur. (The high energy KLL peak of sulphur at 2117 eV is too insensitive for surface layer detection [33].) Another possibility

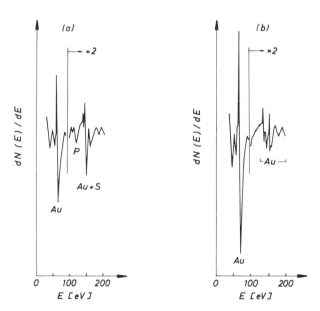

Fig. 18. Low energy Au Auger spectrum. (a) Surface contaminated with S (150 eV) and P (120 eV); (b) clean surface.

would be the reduction of the primary energy below the N_5 level of gold (334 eV), e.g. 300 eV, which will still be sufficient for the ionization of the sulphur L_3 level (164 eV) (compare Table 1). The presence of heavier elements is, of course, more easily recognized because of their numerous peaks, some of which will not overlap with other elements. In general, high energy peaks are more useful for unambiguous qualitative determination. Because they are less sensitive compared with low energy peaks (Fig. 22), the latter are more suitable for quantitative work.

(C) RECOGNITION OF THE INFLUENCE OF CHEMICAL BONDING

According to eqn. (3), any chemical bonding results in a shift of the electron levels in the respective element. Thus an influence of the chemical environment on the Auger spectrum is generally expected. In contrast to ESCA [43] (see Sect. 13.A), three electron levels are involved in the Auger process. Therefore, a "chemical shift" is difficult to predict or interpret. It is often masked by a large natural line width (up to 20 eV) and by secondary effects like plasmon losses. If valence bands are involved (WVV or WXV transitions), the Auger spectrum reflects the changes in the electron density distribution which occur during chemical bonding. A well

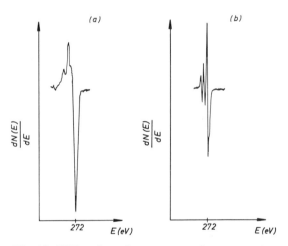

Fig. 19. KLL carbon Auger spectra from (a) carbon monooxide adsorption layer; (b) chromium carbide.

122

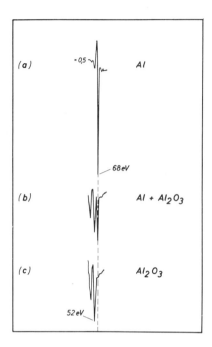

Fig. 20. LVV Auger peaks of Al from (a) metallic Al; (b) both metal and oxide at the surface; (c) oxide only present (from ref. 82).

known example is the carbon KLL Auger peak [76,77]. Metallic carbides show a distinct alteration in the Auger spectrum with respect to graphite or C—O adsorbate carbon. Figure 19 gives an example of its different shapes.

Silicon shows, in addition to a change in shape, a pronounced Auger energy shift due to chemical bonding: its LVV peak at 92 eV is shifted to 87 eV in Si_3N_4 and to 78 eV in SiO_2 [78—80].

Figure 20 shows the low energy Auger spectrum of aluminium in pure metallic form (a) and in Al_2O_3 (c) [81,82]. In this case, the large energy shift of about 15 eV is attributed either to a valence band shift [35] or to so-called cross-transitions between oxygen and metal levels [83,84]. In principle, the electron density distribution in the valence band may be evaluated by application of deconvolution techniques [85,86,259].

The practical analyst will be generally restricted to empirical observations of chemical influence in the Auger spectra [87]. High resolution Auger spectrometry with CMA's or spherical analyzers ($\Delta E/E < 0.5\%$) [224] and careful comparison with internal standards is expected to yield better information on the chemical bond. In situ chemical reactions can be investigated by this method, as recently shown for nitrogen on copper [88]. It should be pointed out, however, that the influence of chemical bonding makes a quantitative analysis rather difficult (see Sect. 8).

8. Quantitative analysis

It was shown very early, that Auger spectroscopy is a quantitative analytical tool [89]. In this section, the basic problems encountered in quantitative evaluation are outlined. Problems arising from in-depth distribution are discussed in Sect. 9.

(A) GENERALIZED QUANTITATIVE MODEL

An exact quantitative Auger analysis requires a unique relation between a measurable Auger electron current and elemental concentration. The two basic problems comprise instrumental parameters (their constancy and calibration) and specimen parameters. The general formulation of the measurable Auger current of WXY-transition of an element i, $I_i(\text{WXY})$ may be given as [20]

$$I_i(\text{WXY}) = \int_\Omega \int_{E_w}^{E_p} \int_0^\infty I(E,z)\sigma_i(E,W)\rho_i(\text{WXY})c_i(z)\,\exp(-z/\lambda)\,\mathrm{d}\Omega\,\mathrm{d}E\,\mathrm{d}z \tag{10}$$

where Ω is the spherical angle of Auger emission, E_p the primary electron energy, E_w the energy of W ionization level, $I(E,z)$ the excitation current in depth z at energy E, $\sigma_i(E,W)$ the cross-section of W level ionization, $\rho_i(\text{WXY})$ = probability of a WXY Auger transition, $c_i(z)$ the concentration in depth z, λ the mean escape depth of Auger electrons which is dependent on Auger energy and matrix [Fig. 2 and eqn. (1)]. In practice, λ will stand for an effective mean escape depth which takes into account the entry angle into the analyzer with respect to the normal on the sample surface, $\varphi : \lambda \rightarrow \lambda \cos \varphi \sim \lambda \times 0.74$ since $\varphi = 42.3°$ for a CMA analyzer [21,66,90] in the arrangement of Fig. 10.

The total excitation current, $I(E, z)$, is composed of the primary

124

electron beam current, I_p, and a contribution caused by high energy primary electrons which are scattered back and possess enough energy for ionization in the escape depth region [32]. This back-scattering electron current, $I_B(E, z)$, leads to an enhancement of the Auger yield and may be taken into account by a term defined by

$$I(E, z) = I_p + I_B(E, z) = I_p r(E, z) \tag{11}$$

where I_p is the primary electron beam current and r is the back-scattering factor. The latter is mainly dependent on the energy distribution of the back-scattered electrons with respect to the ionization energy E_w [32,37,38,40]. Figure 21 gives an example of the back-scattering factor variation with the reduced energy $U = E_p/E_w$ for elements with different Z [37].

With eqn. (11), one can account for the energy integral in eqn. (10) and by introducing the total analyzer transmission Γ one obtains

$$I_i(\text{WXY}) = I_p \Gamma \sigma_i(E_p, W) \rho_i(\text{WXY}) \int_0^{z_e \rightarrow \infty} r(z) c_i(z) \exp(-z/\lambda)\, \mathrm{d}z \tag{12}$$

where z_e denotes the excitation depth of the primary electrons. Since, usually $(E_p \geqslant 3\ \text{keV})\ z_e \gg \lambda$, $z_e \rightarrow \infty$ will be a reasonable approximation.

Integration of eqn. (12) comprises problems of depth distribution as discussed in Sect. 9.

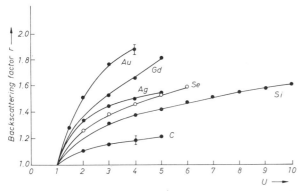

Fig. 21. Measured back-scattering factors, r, as a function of the reduced primary energy $U = E_p/E_W$ for different elements (from ref. 37).

References pp. 166—172

For a homogenous composition within the excitation volume $(c_i(z) = \text{const.})$ it follows [15,23]

$$I_i(WXY) = I_p\Gamma\sigma_i(E_p, W)\rho_i(WXY)rc_i\lambda \tag{13}$$

A calculation of the concentration after eqn. (13) is possible if the ionization cross-section, $\rho_i(E_p, W)$, the transition probability, $\rho_i(WXY)$, the back-scattering factor, r, and the escape depth, λ, are available [200]. Since these parameters are not generally known with sufficient accuracy and measurement of absolute Auger currents is rather difficult, calibration standards (St) of known composition (and of the same surface roughness as the sample) must be used. In this case, the concentration, c_i, is obtained from the intensity ratio of the sample (I_i) to the standard (I_{St}) according to eqn. (13).

$$c_i = c_{St} \cdot \frac{r_i\lambda_iI_i}{r_{St}\lambda_iI_{St}} \tag{14}$$

In general, peak-to-peak heights of the derivative spectrum are used as Auger intensities.

(B) QUANTITATION WITH ELEMENTAL SENSITIVITY FACTORS (FIRST-ORDER APPROXIMATION)

If standard specimens are used with a composition similar to the sample, the escape depths and the back-scattering contributions are the same, giving

$$c_i = c_{St} \frac{I_i}{I_{St}} \tag{15}$$

If no standard is available, an approximate quantification method is frequently applied which is based on the relative elemental sensitivities in the $dN(E)/dE$ spectra [20,90,91]: from eqn. (13), a sensitivity factor $S_i = \sigma_i\rho_ir\lambda$ can be defined so that

$$I_i = I_p\Gamma S_ic_i \tag{16}$$

The total concentration, c, of a sample consisting of m elements can now be expressed as

$$c = \sum_{j=1}^{m} c_j = \frac{1}{I_p\Gamma} \sum_{j=1}^{m} \frac{I_j}{S_j} \tag{17}$$

126

The mole fraction X_i of each element is then given by

$$X_i = \frac{c_i}{c} = \frac{I_i/S_i}{\sum\limits_{j=1}^{m} I_j/S_j} \qquad (18)$$

Since only the ratios S_i/S_j occur in eqn. (18), these relative sensitivity factors can be used instead of the absolute sensitivities defined by eqn. (17). Usually, the relative sensitivities are given with respect to pure silver, which is the most sensitive element in AES. Figure 22 shows the relative elemental sensitivities obtained using the peak-to-peak heights in the $dN(E)/dE$ Auger spectra of the most prominent transitions [33].

Applying eqns. (14)—(18), one must keep in mind the basic approximations made. It is important that the APPH's are really proportional to the Auger current. This can be readily assumed if the shape of the Auger signal is independent of composition. Furthermore, the atomic density is assumed constant for all cases. If this latter condition is violated (e.g. by comparing metals and oxides) c_i must be expressed as $X_i\eta_i$, where η_i is the density in mole per unit volume [16].

Application of eqn. (18) can therefore only be regarded as a semi-quantitative method, that means a first-order approximation with an

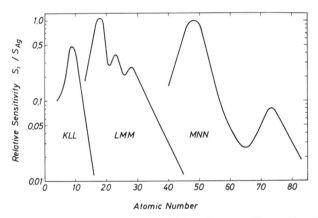

Fig. 22. Relative Auger sensitivity factors ($S_{Ag} = 1$) of the elements at $E_p = 5$ keV (from ref. 33).

TABLE 2

Comparison of X-ray fluorescence (XFA) and AES analysis using elemental sensitivity factors of a sputter-cleaned stainless steel surface (AISI 3162) [70]

	Element			
	Fe	Cr	Ni	Mo
XFA (wt.%)	63.9	17.7	13.3	2.8
AES (wt.%)	65	16	12	6
Auger transition	$L_3M_4M_4$	$L_3M_2M_4$	$L_3M_4M_4$	$M_5N_2N_4$
Auger energy [eV]	703	529	848	186

expected accuracy of $<\pm30\%$ [15,90]. However, applied to the analysis of alloys with similar metallic elements, e.g. stainless steels, a much better accuracy can be obtained. An example for NiCr steel analysis is given in Table 2 [82]. After the removal of the oxide layer by argon ion sputtering (see Sect. 9.B), the Auger peak-to-peak heights of the detected elements Fe, Cr, Ni, Mo have been evaluated according to eqn. (18) with the elemental sensitivities from ref. 33. Comparison of the results with those obtained by X-ray analysis support the applicability of eqn. (18) for Fe, Cr and Ni, but not for Mo. The reason is obvious: the densities of the major constituents and the alloy are similar and the Auger transitions are of the same type for Fe, Cr and Ni. On the other hand, the density of pure Mo is higher, and a MNN Auger transition at lower energies is used for the calculation. (An additional effect due to the low sputtering yield of Mo, which leads to sputtering-induced Mo enrichment in the surface layer, is discussed in Sect. 10.D).

(C) LIMITS OF DETECTION

The Auger electron signal is always superimposed on a large background current which consists mainly of the inelastic scattered primary beam electrons [16,17,26] (see Sect. 5.B). The principal limit of detection is therefore given by shot noise [16,51] according to eqn. (5). For a band pass analyzer as the CMA (see Sect. 5.B), the collected current depends on the magnitude of the energy window (i.e. the absolute energy resolution ΔE). If we assume, as typical values, a CMA resolution of 1% ($\Delta E = 10$ eV at $E_A = 1000$ eV), a bandwidth of 0.1 s^{-1}, a uniform secondary electron current of the order of magnitude of the primary current $I_p \sim 10^{-5}$ A at $E_p \sim$ 3 keV, eqn. (5) with $I \sim I_p(\Delta E/E_p)$ gives $I_N \sim 10^{-13}$ A.

According to a typical Auger yield of $I_A \sim 10^{-4} I_p$ [mainly determined by the respective cross-sections of ionization and Auger transitions, see eqn. (13) in Sect. 8.B], a typical signal-to-noise ratio $I_A/I_N \sim 10^4$ is obtained. This corresponds to about 100 ppma. In this rough estimation, an ideal matching of the natural Auger line width and analyzer energy window is assumed and the analyzer transmission is not respected. More detailed signal-to-noise figure calculations are given in refs. 16, 51. High energy resolution desirable to reveal the fine structure of Auger spectra reduces the detection sensitivity. Large data acquisition times can improve the detection limit according to eqn. (5). However, together with additional noise sources from the analyzer, electron multiplier and amplifier equipment, a typical detection limit of about 1000 ppma is obtained in practice for almost all elements $(Z \geqslant 3)$, and some hundred ppma's for very favourable elements such as silver or sulphur (see Fig. 22).

It must be emphasized that the detection limit defined here is confined to the information depth of the Auger electrons, i.e. a few atomic layers (see Sect. 3). That means that the analysis volume [82,96] which contributes to an Auger signal may be very small. If we assume a detection limit of 1000 ppma for a typical analysis volume of about 10^{-14} cm^3, the "absolute" detection limit is about 10^{-16} g. This shows that Auger spectroscopy may be useful in microanalysis (see Sect. 12.A) and for trace analysis if an enrichment of solute elements in the surface layer of a sample is possible, e.g. by surface segregation (Sect. 9.A).

9. In-depth distribution of composition

(A) THIN SURFACE LAYERS (< 2 nm)

The limitation of the information depth in AES leads to a strong dependence of the output on the depth distribution of the detected species on an atomic scale [229]. Therefore, a correction to the quantitative evaluation given in Sect. 8.B is necessary, at least if a composition change occurs within the respective escape depth of the Auger electrons. Even in the case of negligible matrix effect on backscattering and escape depths, one has to solve the remaining integral term in eqn. (12)

$$I_i = \alpha_i \int_0^{z_e \to \infty} c_i(z)\, \exp(-z/\lambda_i)\, dz \qquad (19)$$

where $\alpha_i = I_p \Gamma \sigma_i(E_p, W) \rho_i(WXY) r(z) = I_p \Gamma S_i/\lambda = \text{const.}$ is a specific Auger yield factor in the respective experiment.

The unambiguous calculation of $c_i(z)$ according to eqn. (19) is only possible if we may assume a certain shape of the depth distribution. Figure 23 shows schematically how the same Auger electron intensity can be obtained for quite different surface compositions depending on the occurrence of the detected element only in the first few atomic layers or in a homogenous depth distribution (as discussed in Sect. 8.A). In the case of adsorption or segregation surface layers, we may assume a constant mole fraction X_A of an element A up to the depth z_1 and $X_A = 0$ for $z > z_1$ in a binary system of elements A and B (compare Fig. 24). From the integration of eqn. (19) we get

$$I_A = \alpha_A \lambda_A X_A [1 - \exp(-z_1/\lambda_A)] \qquad (20a)$$

where $\alpha_A \lambda_A = I_A^0$, the intensity of a bulk elemental standard, follows from $X_A = 1, z_1 \to \infty$.

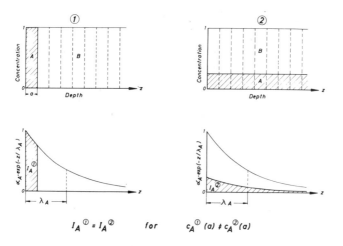

Fig. 23. Visualization of the effect of escape depth, λ_A ($=\lambda_B$), on the Auger yield, I_A, for different in-depth distribution of composition, after eqn. (19) [82].

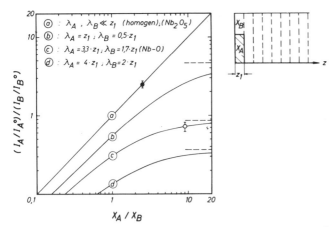

Fig. 24. Calculated normalized Auger intensity ratio $(I_A/I_A^0)/(I_B/I_B^0)$ [eqn. (21)] as a function of the X_A/X_B for different ratios of escape depths λ_A, λ_B and film thickness z_1. ● and ○ denote measured points for Nb_2O_5 and Nb—O segregation layer (compare Fig. 25) [82].

For element B, we have to add the contribution beyond z_1 which gives

$$I_B = \alpha_B \lambda_B \{X_B[1 - \exp(-z_1/\lambda_B)] + \exp(-z_1/\lambda_B)\} \qquad (20b)$$

With the intensities of bulk elemental standards, I_A^0 and I_B^0 $(=\alpha_B\lambda_B)$, combination of eqns. (20a) and (20b) yields the ratio X_A/X_B as a function of the measured Auger intensities.

$$\frac{X_A}{X_B} = \left\{\frac{I_B}{I_A}\frac{I_A^0}{I_B^0}[1 - \exp(-z_1/\lambda_A)] - \exp(-z_1/\lambda_B)\right\}^{-1} \qquad (21)$$

If no standard measurements are available, the ratio I_A^0/I_B^0 may be replaced by the relative elemental sensitivity factor S_A/S_B defined in eqn. (16).

A similar relation was derived by Seah [66] for the case of thin evaporation layers. The ratio of the depth z_1 and the mean escape depths λ_A and λ_B are the relevant parameters in eqn. (21) as seen in Fig. 24. In the limiting case of a thick layer of A, $z_1 \gg \lambda_A$, λ_B we arrive, of course, at eqn. (18).

An important consequence of eqn. (21) is the fact that a linear relation between X_A and the intensity ratio I_A/I_B is only obtained for

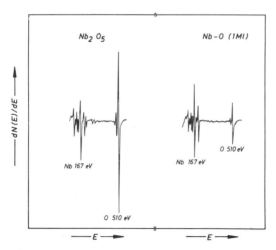

Fig. 25. Auger spectra for electrochemically produced Nb_2O_5 and for a mono-layer of oxygen segregated at the surface of niobium [82].

very small values of X_A or for $\lambda_B \gg z_1$. If these conditions are not valid, as, for example, in the case of adsorption or segregation of A in the monolayer regime at higher coverage, eqn. (21) must be applied [82,92]. For example, in surface segregation studies of oxygen in niobium (bulk concentration <1 at.%), a maximum oxygen signal intensity as shown in the Auger spectrum of the right-hand side of Fig. 25 was obtained. For comparison, the Auger spectrum of an electrolytically produced Nb_2O_5 layer (400 Å thick-ness) is also depicted. It may be used for a calibration of the Nb—O system according to eqns. (15) and (18), which yield about 30 at.% oxygen. On the other hand, from adsorption experiments with oxygen and from in-depth composition profiles [69] we should expect a 100% oxygen coverage restricted to the first monolayer for the Nb—O peak-to-peak ratio in Fig. 25. This discrepancy is explained by the fact that eqns. (15) and (18) are only applicable for homogenous composition, as for Nb_2O_5. For the case of thin (adsorption or segregation) layers, however, eqn. (21) must be applied. Taking as a rough estimate for the thickness of the segrega-tion layer $z_1 \sim 3$ Å, λ_{oxygen} (510 eV) ~ 10 Å and λ_{Nb} (167 eV) $\sim$ 5 Å, with the sensitivity ratio calculated from eqn. (18), we get an oxygen coverage of about 0.9 monolayer equivalents for the maxi-mum segregation level (Nb—O, Fig. 25) [69,92]. The points for

Nb_2O_5 and Nb—O are shown in Fig. 24. It must be noted that this example is only a first approximation to quantification, because very different matrices (metal oxide and metal) are compared. Therefore we should expect an influence on the relative elemental sensitivities due to the different back-scattering contributions and the different escape depths. These must be considered in a more accurate approach to quantification.

With the procedure outlined above, it is possible to evaluate the thickness of an overlayer if the (bulk) sensitivity factors and the escape depths are known [93,94]. This can be applied in the determination of the thickness of thin evaporation layers or in the measurement of escape depths [66,95,251,252].

Variation of the excitation intensity with depth by use of different angles of incidence of the primary beam can give at least qualitative information about whether an element is concentrated at the surface, because the surface atoms are preferentially excited at near glancing incidence [96]. Variation of the effective escape depth by tilting the sample before the analyzer also allows an estimate to be made of the distribution of an element within the first few surface layers [19].

A very useful method in AES has proved to be the simultaneous observation of two Auger signals of the same element with widely different energies, and hence, escape depths [82,97]. This is possible for the heavier elements $(Z > 10)$ with Auger transitions in the different excitation levels (e.g. LMM and MNN in copper). An example of the build-up of a segregation layer of Sn (of mono-atomic thickness) at the (111) surface of copper containing 0.5 at.% Sn [98] is shown in Fig. 26. The fact that with increasing Sn coverage the low energy (60 eV, $\lambda \sim 3$ Å) Cu signal decreases with a greater slope than the high energy signal (920 eV, $\lambda \sim 12$ Å) [21] is a clear indication that Sn is enriched in a thin overlayer. Note that we must refer to the escape depth of the Cu Auger electrons in Sn. A calculation according to eqn. (21) assuming a tin segregation confined to the first monolayer gives for, $I_{Sn} = 60$ mm (Fig. 26), $X_{Sn} \sim$ 0.3 which is an agreement with LEED investigations performed with this system [99,253].

The last example points out the possibility of using AES even for trace analysis: if the segregation equilibrium constant between surface and bulk concentration is known for a solute element [69,99], a quantitative surface analysis permits the determination of

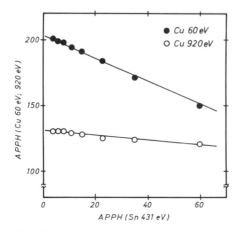

Fig. 26. Auger peak-to-peak heights (APPH) of Cu (920 eV) and Cu (60 eV) compared with Sn (431 eV) during segregation of Sn at a Cu(111) surface [82].

the bulk concentration in the case of attainment of equilibrium at elevated temperatures. Enrichment factors of $\geqslant 10^3$, as for <0.1 at.% oxygen in niobium [69,92], allow the determination of bulk concentrations in the ppm range.

Generally, the accuracy of depth information depending on the mean escape depth of the respective Auger electrons is rather limited. In addition to the fact that this quantity is only known exactly for special cases [21], quantitation is only possible if the integral term in eqn. (12) can be solved, that means $c(z)$ is predictable. Furthermore, this method is limited to depths of the order of a few λ's (<50 Å). These disadvantages are circumvented by the combination of AES with sputter erosion of the sample surface, a method which is generally applicable to the determination of concentration—depth profiles [23,67,100].

(B) DEPTH PROFILING BY SPUTTERING

The depth distribution of chemical composition can be obtained by bombarding the surface with energetic ions (>100 eV). Due to the successive sputtering of surface layers, deeper layers are laid free and can be readily analyzed by a surface-sensitive method such as AES. In general, noble gas ions (Ar^+) are used because disturbing chemical reactions should be avoided. A prerequisite for the mea-

134

surement of depth profiles is a constant erosion rate within the analyzed area [101]. In AES with CMA equipment, this condition is easily achieved because the analyzed area (typically $\leqslant 50\,\mu$m in diameter) is much smaller than the primary ion beam diameter (typically $\geqslant 2$ mm). The basic problem in the evaluation of the sputtering profiles lies in the transformation of the Auger signal intensities I_i (peak-to-peak heights) versus sputtering time t, $I_i = f(t)$ to the elemental concentration c_i as a function of the eroded layer depth z, $c_i = f(z)$ [82,101]. The problems encountered are shown schematically in Fig. 27. The most simple method is to determine $c_i = f(I_i)$ according to the first-order quantification procedure outlined in Sect. 8.B, eqn. (18), and to get $z = f(t)$ with the assumption of linear dependence between sputtering time and eroded depth, $z = \dot{z}t$ [101,102]. The erosion rate $\dot{z}$ can be determined either by knowledge of the primary ion current density and the sputtering yield [102] of the system under investigation, or by calibration with similar layers whose thickness has been measured by other means. An elegant method of depth scale evaluation is simultaneous monitoring of the characteristic X-ray emission [228]. Numerous examples of concentration—depth profiles are given in the literature [103,104].

In spite of the apparent success in obtaining composition profiles, however, it must be pointed out that the reliability of concentration—depth profiles depends on quite a number of effects which have

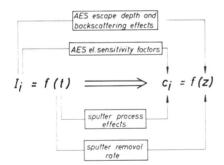

Fig. 27. Problems encountered in the evaluation of concentration (c_i) versus depth (z) profiles from Auger intensity (I_i) versus sputtering time (t) profiles. First-order approximation is possible using elemental sensitivity factors and constant sputter removal rate. A more reliable analysis must account for electron excitation and escape depth variations due to sample alteration during the sputtering process.

to be carefully controlled. They may be divided into instrumental factors and factors depending on the sputtering process itself [100,101,104].

The instrumental factors (e.g. primary ion beam homogeneity and stability) can be minimized by a proper setting up of the apparatus. The role of the complex nature of the sputtering process [100,102,105] in depth profiling is, however, far less accessible. A careful evaluation must take into account effects such as changes in erosion rate and surface composition caused by different sputtering yields of the sample constituents ("preferential sputtering") [106–108] (see Sect. 10.D), changes in in-depth concentration by ion implantation ("knock-on" effect and ion mixing [109–111]), ion bombardment-induced diffusion and chemical reactions [112], changes in the surface topography due to the dependence of ion sputtering yields on crystal orientation and structural imperfections [102,113], and the original surface roughness [110,114] etc. Super-imposed on these effects is the statistical nature of the sputtering process: in a first-order approximation, it may be assumed that any part of a deeper layer which has become a surface layer during sputtering will be eroded with the same probability as the original first layer [115].

All the processes mentioned above will operate simultaneously in surface erosion by ion bombardment. Their relative influence on depth profiling depends strongly on the sample under investigation [104]. Their common feature is a distortion of a direct similarity between the measured sputtering profile and the "real" concentration profile in the sample. In general, the measured profile is broadened with respect to the real profile [101,103,116,117]. Figure 28 [114] shows, as a typical example, the AES sputtering profile of a multilayer sandwich structure which consists of ten Ni/Cr layers alternatively evaporated on a glass substrate. Sputtering was performed with 1 keV argon ions. The difference between Fig. 28(a) and (b) is due to the original surface roughness. In Fig. 28(a), the layers were evaporated on a polished glass substrate with a mean roughness smaller than about 10 nm whereas in Fig. 28(b), the glass substrate had a mean roughness of about 1 μm. The degradation of the shape of the "real" rectangular profiles increases with sputter depth in both cases but is much more pronounced in case Fig. 28(b) with greater roughness. Note that the relative sensitivities of Cr and Ni depend also on surface roughness [118].

136

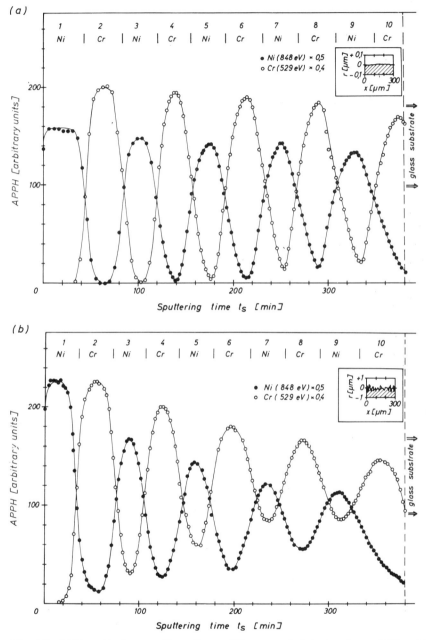

Fig. 28. AES sputtering profiles of a multilayer sandwich structure of Ni and Cr layers alternately evaporated on a glass substrate with an original surface roughness of (a) ⩽ 10 nm; (b) ~1 μm [114]. Thickness of each layer = 37 nm.

The observed sputter broadening can be described by a convolution integral of the form [116]

$$I_i^n(z_1) = \int\limits_{-\infty}^{+\infty} c_i(z_1 - z)\, g(z_1 - z)\, dz \tag{22}$$

where I_i^n is a normalized intensity of element i at depth z_1 which depends on the contributions of local concentrations in the depth $z_1 - z$. The function g defines the depth resolution. Although mathematically complicated, in principle a deconvolution is possible if $g(z)$ is known. Experimentally, the resolution function is directly obtained if sputter profiling is applied to an infinitesimally thin sandwich layer. The integrated form is obtained if sputtering proceeds through a step function like profile of an ideally smooth interface and the intensity response is observed [101,107,108]. In the latter way, an intensity profile in the shape of a Gaussian integral function or error function is generally obtained. Therefore, we only need one parameter, Δz, for the definition of depth resolution as visualized (Fig. 29). We have chosen the most usual Δz definition according to the 2σ value [101,103,104,107,108] where σ is the standard deviation of the respective error function as determined by the difference

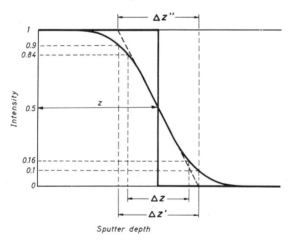

Fig. 29. Definition of depth resolution Δz. For an error function-like profile, $\Delta z = 2\sigma$ where σ is the standard deviation. Other definitions sometimes used are: $\Delta z'$ (0.1—0.9) = 2.564σ, $\Delta z''$ (inverse max. slope) = 2.507σ ($\Delta z'''$ taken at FWHM of derivative = $2.355\,\sigma$).

138

of the z values between 84% and 16% of the normalized intensity. Alternatively, other definitions are sometimes used, such as the 90% to 10% value [119,120], the inverse maximum slope, and the full width at half maximum of the derivative Gaussian distribution [116] (Fig. 29).

The experimentally determined depth resolution term Δz is composed of the effects mentioned above, which depend on the primary ion beam and the chemical and structural nature of the sample. If the individual contributions are independent of each other and may be approximated by an error function, it can be shown mathematically that the total measured Δz is given by the root mean square of all the individual Δz_j's [82,104].

$$\Delta z = [\sum_{j=1}^{n} (\Delta z_j)]^{1/2} \qquad (23)$$

A general prediction of the different magnitudes of the Δz_j contributions is rather difficult because of the complexity of the sputtering process. In metallic samples, cone formation is often encountered [100,102,105] during sputtering with high energy argon ions. Sputtering with low energy ions (500 eV to 1 keV) generally leads to an improvement in depth resolution [100,110]. Reactive primary ions (e.g. oxygen ions) can be applied to avoid cone formation [113,121].

The role of atomic transport in the surface region is expected to be of decisive importance: if diffusion in the bulk perpendicular to the surface is prevalent, the composition versus depth profile will change during sputter erosion; if diffusion across the surface prevails, the depth resolution may be improved because microroughening due to the sputter statistics is more or less outweighed [104,114].

A simple sputter erosion model has been described by Benninghoven [115]. The basic assumptions are (a) only particles in the topmost layer are removed and (b) the sputter erosion rate is constant for the successive layers. If τ is the mean erosion time for one layer with a thickness a

$$\tau = \frac{N_0 e}{j_p \gamma} \qquad (24)$$

where N_0 is the number of atoms in a layer of thickness a, e is the elementary charge, j_p is the primary ion current density, γ is the sputtering yield. Each layer is sputtered away following an exponential law with the normalized sputtering time t/τ.

$$\theta_n(t) = \theta(0)\exp(-t/\tau) \tag{25}$$

Since that part of the next layer which has become a surface layer after a certain sputtering time is eroded by the same law, the relative part $\theta_n(t)$ of the nth layer which is a surface layer at a sputtering time t is given by [82,101,115]

$$\theta_n(t) = \frac{(t/\tau)^{n-1}}{(n-1)!}\exp(-t/\tau) \tag{26}$$

This is the basic equation of the so-called "sequential layer sputtering model" [104,114,117].

The total momentary surface consists of the sum over all $\theta_n(t)$. If we refer to the sputtering through a sharp interface ($\theta_n = 0$ for $n > N$) the shape of the profile is given by

$$\theta(t) = \sum_{n=1}^{N} \theta_n(t) \tag{27}$$

An example of calculated profiles for $N = 20$ and $N = 40$ single layers is shown in Fig. 30 [101]. Although the profiles are slightly

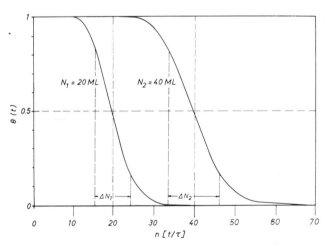

Fig. 30. Example of two calculated profiles for sharp interfaces at 20 and 40 monolayer thickness according to eqn. (27) [101].

asymmetric, they resemble that of Fig. 29. It can be shown mathematically that the integral over the Poisson distribution represented by eqn. (27) approaches an error function (Fig. 29) for large N (e.g. $N > 10$) [122]. Hence the standard deviation σ is given by the square root of $t/\tau = N$. According to the 2σ definition of depth resolution (Fig. 29), we get

$$2\sigma = \Delta N = 2N^{1/2} \tag{28}$$

In terms of the total layer thickness $N = z/a$, the absolute depth resolution Δz_s following from the sequential layer sputtering model is dependent on the total sputter depth according to

$$\Delta z_s = 2(az)^{1/2} \tag{29a}$$

The relative depth resolution is given by

$$\Delta z_s/z = 2(a/z)^{1/2} \tag{29b}$$

Evidence of eqns. (29) has been found in quite a number of metallic systems as shown in Fig. 31 [82,104] with the parameter a of the order of a monolayer thickness. It must be pointed out, however, that the prerequisite for the applicability of the sequential layer

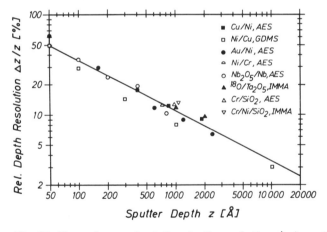

Fig. 31. Dependence of relative depth resolution $\Delta z/z$ on the depth removed by sputtering compiled for various systems (included are glow discharge mass spectrometry, GDMS, and ion microprobe mass analysis, IMMA, results). ■, ref. 116; □, ref. 119; ●, ref. 110; ◓, ref. 124; ○, ref. 101; ▲, ref. 123; △, ref. 124; ▽, ref. 125. The straight line corresponds to eqn. (29b) with $a = 0.29$ nm [104].

sputtering model, namely a constant erosion rate for small and large surface layer fractions, will not be achieved. Therefore the prediction of the model tends to give too large Δz values for a taken as a mono-layer thickness. On the other hand, all the effects discussed above will operate, and will superpose according to eqn. (23). Therefore a is merely a fitting parameter, which may be of subatomic size as expected if atomic transport across the surface will suppress micro-roughening [104,114]. The general applicability of the sequential layer model depends on the question of to what extent the instrumental factors and the atomic transport processes in the sample can be neglected. Unfortunately, they can hardly be predicted at present. Atomic transport improved depth resolution will be more pronounced in semiconductor systems [126] and may even lead to constant depth resolution at larger sputter depth [127].

At least one of the most fundamental contributions to Δz in AES sputter profiling which may be estimated is the effective escape depth of Auger electrons [128]. For an ideal microsectioning, we should expect an Auger intensity behaviour by approaching the sharp interface according to eqn. (20a) as shown in Fig. 32 for differ-ent ratios of λ to the total depth. Obviously, the relative depth resolution is improved if λ is small compared with the total depth.

Applying the sequential layer sputtering model, the actual Auger signal intensity of an element i, I_i, at sputtering time t is composed

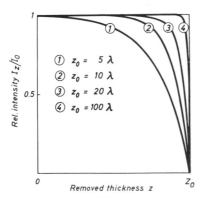

Fig. 32. Auger intensity versus sputtering depth profile expected for ideal micro-sectioning when a sharp interface of z_0 is approached; from eqn (20a), it follows, with $z_1 = z - z_0$.

$$I_z = I_0 \left[1 - \exp\left(\frac{z_0 - z}{\lambda}\right) \right]$$

142

of the contribution of all layer fractions. Analogous to eqn. (27), we may write

$$I_i(t) = \sum_{n=1}^{N} I_{i,n}(t) \tag{30}$$

where $I_{i,n}(t)$ is given by the $\theta_n(t)$ term in eqn. (26) and the contribution of all layers below $\theta_n(t)$ following eqn. (19).

$$I_{i,n}(t) = I_{i,0} \sum_{l=0}^{N-n} X_{i,n+1} \theta_n(t) \exp(-la/\lambda_i) \tag{31}$$

$I_{i,0}$ is a normalization factor for the Auger signal intensity for $l = 0$; $x_{i,n+1}$ is the atomic fraction of component i in the layer n + 2. Note that the Auger yield integral in eqn. (19) is expressed by a sum over the subsequent layers ($z \rightarrow la$) of thickness a.

Results of the sputtering profile calculation according to eqns. (30) and (31) for a thin film of 50 layers are shown in Fig. 33. For increasing λ_i, the profile broadening increases, and the apparent interface location at $I_i = 0.5\, I_{i,0}$ is shifted away from the interface by an amount of about 0.7 λ_i. The influence of the escape depth on the depth dependence of Δz according to the sequential layer sputtering model is shown in Fig. 34 [82]. For increased surface micro-roughening, the λ effect becomes negligible [114].

In spite of its strong idealization, the sequential layer model has

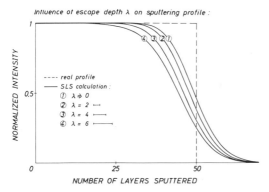

Fig. 33. Sputtering profile calculation of 50 layers for different values of the escape depth λ (in monolayers) from eqn. 31.

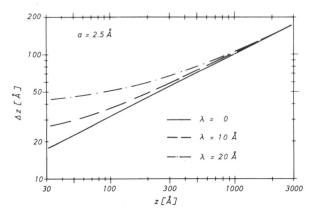

Fig. 34. Depth resolution, Δz, as a function of sputter depth, z, for different values of λ, as expected from eqn. (23) [104]. $\Delta z = (\Delta z_\lambda^2 + \Delta z_s^2)^{1/2}$ with $\Delta z_\lambda = 2\lambda$.

been applied with considerable success [101,104,114,117]. A comparison between profile calculation using eqns. (30) and (31) and measurement is depicted in Fig. 35 for a multilayer sandwich sample of Ni/Cr layers of 115 Å thickness [114]. Agreement is good up to the 13th layer. At larger depths, increased oxygen content in the layer or atomic transport induced by electron beam heating (see

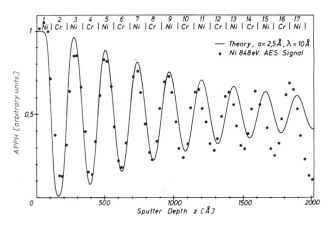

Fig. 35. Comparison between measured sputtering profile (●) for Ni in a Ni/Cr sandwich multilayer sample and calculation from eqns. (30) and (31) [114]. The thickness of each layer is 155 Å.

144

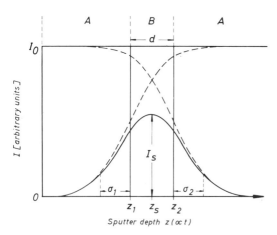

Fig. 36. Expected sputtering profile $I(z)$ for a single sandwich layer of B in matrix A with thickness d from superposition of two error functions (standard deviations σ_1, σ_2) according to eqn. (32).

Sect. 10.C) may have been responsible for the observed deviations.

An example of the composition profile broadening due to sputtering is the measured profile expected from a single sandwich layer of a component B with thickness d in matrix A as shown in Fig. 36 [117]. It is a superposition of two error functions (Figs. 29 and 30). The normalized intensity $I(z)/I_0$ has the form [117]

$$\frac{I(z)}{I_0} = \frac{1}{2}\left[\, \text{erf}\left(\frac{z - z_1}{\sqrt{2}\sigma_1}\right) - \text{erf}\left(\frac{z - z_1 - d}{\sqrt{2}\sigma_2}\right)\right] \tag{32}$$

where z_1 and z_2 denote the location of the interfaces, and σ_1 and σ_2 the standard deviations of the respective error functions. Approximately, $\sigma_1 \sim \sigma_2$ for $d \gg z_1, z_2$. With the definitions of depth resolution (Fig. 29) $\Delta z = 2\sigma$, the maximum of the sputter profile $I(z) = I_s$ is reached at $z_s = z_1 + (d/2)$. For this value, eqn. (32) gives

$$\frac{I_s}{I_0} = \text{erf}\left(\frac{d}{\sqrt{2}\,\Delta z}\right) \tag{33}$$

Equation (33) allows the depth resolution to be determined if the sandwich layer thickness, d, is known (and vice versa), when I_s/I_0 is taken from the measured profile (Fig. 37).

An example of the application of eqn. (33) is shown in Fig. 38

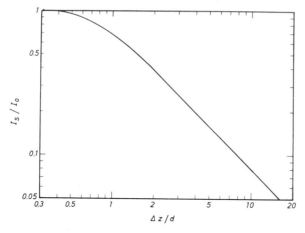

Fig. 37. Relation between normalized maximum intensity I_s/I_0 from Fig. 36 and the ratio between depth resolution and layer thickness, $\Delta z/d$, from eqn. (33).

for the case of two Cr sandwich layers in a Ni matrix which were evaporated on a smooth glass substrate [117]. The theoretical curves have been calculated from eqn. (32), $\sigma_1 = \sigma_2 = \Delta z_s/2$, and eqn. (29a), $a = 0.25$ nm. Measured and calculated parameters are compiled in

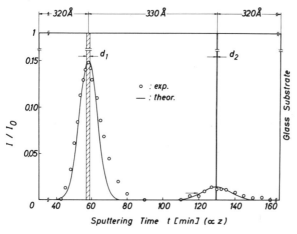

Fig. 38. Measured AES sputtering profiles of two Cr sandwich layers in a Ni matrix (○) and the corresponding curves calculated from eqns. (29a) and (32). The relevant parameters are listed in Table 3.

146

TABLE 3

Comparison of measured and calculated values for the AES sputtering profile of two Cr layers subsequently evaporated on Ni (Fig. 38)

z_1 = thickness of Ni layer on the Cr layers; d_{ev} = Cr layer thickness determined by a microbalance during evaporation; I_s/I_0 = normalized maximum intensity (I_0 = intensity for thick Cr layer); $\Delta z/d$ and Δz_{ev} calculated from eqn. (33) using d_{ev}; d calculated using Δz_s from eqn. (29a) with $a = 0.25$ nm.

Cr layer no.	z_1 [nm]	d_{ev} [nm]	d [nm]	I_s/I_0	$\Delta z/d$	Δz_{ev} [nm]	Δz_s [nm]
1	32	1.03	1.1	0.15	5.2	5.4	5.7
2	65	0.21	0.14	0.014	57	12.0	8.1

Table 3. According to Fig. 32, the escape depth influence ($\lambda \sim 1$ nm) is neglected. Table 3 and Fig. 38 show remarkable agreement between measured and calculated values. The theoretical layer thickness $d = 0.14$ nm for the second Cr layer can be interpreted as about 0.6 of a monolayer ($d = 0.25$ nm) (or 0.3 of two monolayers etc.) without affecting the measured profile. This demonstrates the meaning of depth resolution Δz as a depth range, within which any concentration change cannot be resolved. Hence, original interface "roughness" is tolerable if it is small compared with Δz.

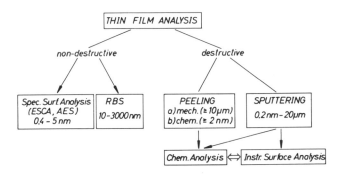

Fig. 39. Schematic diagram of different analytical possibilities to obtain the in-depth distribution of composition in thin films. Spec. analysis means λ-dependent or take-off angle dependent electron energy analysis (see Sect. 9.A). RBS means rutherford back-scattering spectroscopy of high energy ions (see Sect. 13.C). For other instrumental analysis methods, see Sect. 13.

An important consequence of eqn. (33) is the fact that the detection of a thin layer is limited by the depth resolution. Taking a typical value for the detection limit in AES profiling $(I_s/I)_{min} \sim 1\%$, the minimum thickness of a layer to be detected is d_{min} (1%) = 0.0128 Δz [eqn. (33)]. For example, a monolayer of about 2.5 Å thickness can only be detected for $\Delta z \leqslant 200$ Å.

Depth distribution of composition and sputter profiling are interlinked with quantification of AES. A combination of several surface analysis methods (see Sect. 13) together with other microanalytical techniques [129] such as electrochemical peeling and subsequent analysis, will be most fruitful in the solution of the outlined problems. Figure 39 shows schematically the different possibilities for obtaining the in-depth distribution of composition in thin films.

10. Error sources

Factors affecting the quantification of Auger analysis, such as the angle of incidence of the primary electrons as well as the take-off angle, back-scattering and chemical shift have already been considered in Sects. 8.B and 9. Recent investigations [130,131] have shown that, in crystalline solids, the Auger emission will be anisotropic for low energies (<300 eV). On the other hand, the low energy Auger lines are superimposed on a steep gradient of the "true" secondary electron distribution (see Sect. 4). It varies largely with the surface state of the sample and thus may appreciably change the peak-to-peak heights in the $dN(E)/dE$ spectrum. The strong dependence of the electron multiplier gain in the lower energy region and its dependence on the ageing of the multiplier (normally used in CMA analyzer) must be considered. With respect to these effects, special care is necessary in the quantification of low energy Auger spectra.

The most important sources of error, however, may arise from the surface morphology and the chemical state of the sample under investigation.

(A) SURFACE MORPHOLOGY

So far, only smooth (polished) surfaces have been considered. In practical work, one is sometimes concerned with very rough surfaces,

148

e.g. fracture surfaces. Here, variations of the incidence angle as well as shadowing effects will influence the Auger yield. In general, surface roughness reduces the Auger output with increasing incidence and/or emission angle [118]. Moreover, the influence of the roughness may alter the relative sensitivity factors [114]. There is little systematic work in this field at present, but deviations of the absolute intensity up to some ten per cent seem to be possible for a roughness in the μm range compared with polished surfaces [62,118]. An example of such a comparison is seen in Fig. 28 in the different Auger peak-to-peak height ratios of Cr and Ni for a rough and a smooth sample surface, at an incidence angle of 60° to the surface normal.

(B) CHARGING OF INSULATORS

Non-conductive specimens pose severe problems since electrons can be trapped and cause charge-up if the secondary electron emission coefficient is less than one. This causes a time-dependent shift of the Auger spectrum. Moreover, the charge-up is generally unstable due to internal breakthroughs after sufficient electron bombardment.

To overcome difficulties with charging, the operating parameters have to be chosen to achieve a secondary electron emission coefficient equal to or larger than unity. This is possibly by variation of the primary energy and the angle of incidence. Medium primary energy (1—1.5 keV) and glancing incidence [17] and low electron density (by defocussing the primary beam) have been found useful for this purpose [79]. If charging cannot be suppressed in this way, a metallic mask at the surface can be added [16]. Another way is to deposit a submonolayer amount of metal on the insulator surface, for example by sputtering a gold target mounted in the vicinity of the sample. Of course, these latter procedures interfere with the analysis but its effect can be controlled.

Migration of ions is often encountered in insulators together with charge-up [112] which can give erroneous results. This is the reason why sodium can hardly be detected in the Auger analysis of glasses [112,124,254].

(C) ELECTRON BEAM-INDUCED ARTIFACTS

One of the basic requirements of a reliable Auger analysis is a vanishing influence of the primary electron beam on surface composition [132,223]. Decomposition [133], electron-induced desorption [134,135,232], deposition [242] and diffusion induced by Joule heating [136] are the most important phenomena frequently observed. Decomposition is expected in ionic compounds such as SiO_2 which is gradually decomposed to elemental Si under electron bombardment [17,133,137,239]. Electron-induced desorption of adsorbates (e.g. oxygen) can be measured by the time dependence of the decreasing adsorbate Auger signal so that the relevant cross-sections can be derived [135]. Out-diffusion enhanced by electron bombardment can be observed for species such as phosphorus [16] or sulphur [126] with rather high diffusivities which tend to segregate at the surface. Even in thin metallic films, heat dissipation may be sufficient to cause diffusion [136,138,255].

Figure 40 shows the change in surface composition of a metallic glass during Auger analysis. After sputtering ceases ($t = 0$), phosphorus is gradually replaced by sulphur (which is present only as an impurity in the bulk) under the Joule heating of the electron beam. The parabolic time dependence of the sulphur signal indicates that diffusion is responsible for the observed process [98,256].

The variety of effects [132—139] due to electron beam—surface

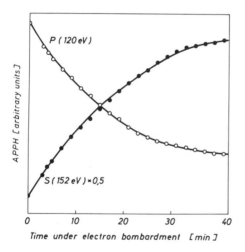

Fig. 40. Alteration of surface concentration of S and P in a metallic glass ($Fe_{32}Ni_{36}Cr_{14}P_{12}B_6$) due to Joule heating induced by the primary electron beam (5 keV, 50 μA).

150

interaction can be minimized by low flux density. With the possible exception of very fast processes, they can be readily observed in prolonged analysis so that extrapolations to zero time is possible. Metallic samples will be less prone to electron-induced composition changes than oxides.

(D) ION BEAM-INDUCED ARTIFACTS

In any case where the contamination layer on a sample surface is not of prime interest, one must generally rely on sputter cleaning to get the prerequisite for a quantitative Auger analysis of the sample. For in-depth analysis, sputtering cannot be circumvented and its effect on depth profiles has been discussed in Sect. 9.B.

Any alteration of the composition of the first few layers due to ion bombardment will show up in AES analysis [100,101,105]. The most prominent effect is selective or preferential sputtering caused by different sputtering yields of the elements [102]. For a binary system of A and B with sputtering yields γ_A and γ_B, a relation between the relative surface composition and (homogenous) bulk composition after a certain sputtering time to reach steady state is given by [106,240]

$$\left(\frac{X_A}{X_B}\right)_{surface} = \frac{\gamma_B}{\gamma_A} \left(\frac{X_A}{X_B}\right)_{bulk} \tag{34}$$

For a precise quantitative analysis, the depth of the altered surface layer also has to be taken into account [111,241]. Further complications arise if the sample is composed of more than two elements or of elements in different binding states, as in multiphase samples. The preferential sputtering effect causes severe problems in the preparation of standards as outlined in Sect. 6.B.

Since sputtering also distorts the local arrangement of the surface atoms and, hence, chemical binding states, quantitative AES may be influenced by peak shape alterations due to chemical shifts (see Sect. 7.C). A typical example is the sputtering of oxides [220,221].

Implantation of primary ions as well as knock-on implantation of target atoms [109,111,226] is generally expected but seems to play only a minor role in AES and sputtering with argon ions below 1 keV energy [109,110], when compared with preferential sputtering.

Enhanced mobility of surface constituents due to radiation

damage of the sample may cause bulk and surface diffusion-dependent artifacts [111,112,222]. Often, surface segregation will provide the driving force for diffusion-assisted enrichment of an element in the surface [69,98,257].

If electron and ion beams are used simultaneously as in sputter profiling (see Sect. 9.B), combined effects of surface decomposition may arise [139,140,220]. Furthermore, using argon ions with energies >1 keV, an increasing effect on the normal Auger spectrum is encountered by ion-induced Auger excitation with somewhat different line shapes [47]. This effect can be used for a proper alignment of electron and ion beam for an optimum depth profiling set-up (see Sect. 9.B).

11. Applications

AES, which is frequently used in conjunction with related methods such as LEED, ESCA, SIMS and ISS, has become one of the most powerful methods of compositional surface characterization. The application potential of AES is schematically shown in Fig. 41. Besides original surfaces, interfaces (e.g. grain boundaries) can be studied if they become a surface by an intergranular fracture. In combination with sputtering, in-depth analysis is possible. Thus, together with a highly focussed scanning electron beam, a three-dimensional analysis can be performed with optimum depth resolu-

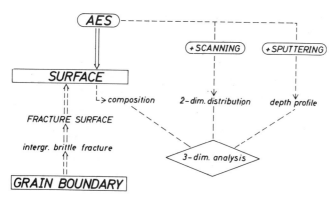

Fig. 41. Principle of the application of AES for localized surface and interface analysis.

152

tion of some monoatomic layers and lateral resolution down to some hundred Ångströms.

Of the numerous applications in many areas of materials science published, some typical examples will show the main fields of interest, which are concerned with surface, interface and in-depth composition as well as with fundamental studies of reactions at surfaces. A recent compilation of AES and other methods applied to surface characterization of a variety of materials is given by Kane and Larabee [141].

(A) SURFACE AND IN-DEPTH COMPOSITION

The detection of surface contamination is one of the major areas of AES [57,142,143]. The processing and manipulation of materials in the surrounding atmosphere and even cleaning procedures give rise to surface contamination. Besides oxygen, carbon, sulphur and chlorine have most frequently been detected at surfaces (see Fig. 11). Fluorine residues are often found in surface layers of metals after etching with hydrofluoric acid. The identification of surface contaminants is supported by in-depth analysis with argon ion sputtering, which shows their confinement to the very first surface layers. Care must be taken, however, to consider contamination within the Auger spectrometer mainly due to CO from the residual gas atmosphere. On the other hand, volatile contaminants and those which are extremely sensitive to electron irradiation-induced desorption cannot be detected by AES.

Surface contamination of carbon has been found to be responsible for increased electrical contact resistance of relay devices [144]. In semiconductor work, carbon has been shown to reduce the photo-yield of GaAs photocathodes [145]. Poor bonding characteristics of gold lead frames to silicon wafers have been ascribed to the presence of thallium at the gold surface [146], the source of which was the gold plating solution applied.

In quality control and in the development of fabrication processes of thin film devices, AES together with composition profiling has become a standard technique. Comparison of the AES results with mechanical, optical or electrical properties can lead to failure recognition and subsequent improvements in processing. A correlation of electrical properties to composition profiles of evaporated thin NiCr films revealed the influence of evaporation parameters on resistivity, contact resistance and temperature coefficient of resistance [124]. Different production techniques (evaporation from a tungsten boat, evaporation with a jumping electron beam, sputter deposition) could be characterized by means of AES depth profiling [147]. Figure 42 shows the difference between a layer produced by evaporation from a tungsten boat and a sputter-deposited layer. Only the latter possessed the electrical parameters expected for a homogenous NiCr (80/20) thin film. Adhesion, friction and wear has been correlated with AES measurements [148,149,235]. Oxidation and corrosion is a broad field of AES application. Passive films at

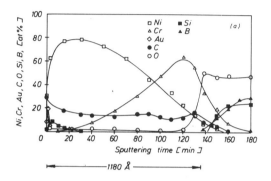

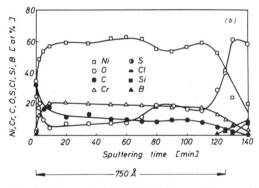

Fig. 42. Example of AES depth profiling applied to the evaluation of the difference of two Ni—Cr evaporation layers produced by (a) evaporation from a tungsten boat; (b) sputter deposition from Ni—Cr 80/20 alloy [in (a), Au has been additionally deposited onto the surface]. First-order quantitation was made using eqn. (18) [147].

stainless steel surfaces have been frequently studied by AES depth profiling [150—152], which revealed the role of Cr enrichment in the formation of these layers. The poisoning of catalysts has been determined by AES [153,154].

AES depth profiling is frequently applied to diffusion studies in metallic thin films [181,182]. Due to the high depth resolution, the useful mean diffusion length can be of the order of a hundred Ångströms thus allowing the determination of interdiffusion constants as low as 10^{-20} cm^2/s. Interdiffusion in Au/Ni [155], Cu/Al [156,238] and Ni/Fe [157] has been studied by AES depth profiles, which enabled a distinction to be made between lattice and grain boundary diffusion [156,157].

(B) FRACTURE SURFACES

The analysis of fracture surfaces by AES is mainly applied to the study of the composition of grain boundaries in metals and ceramics [97,125] (see Fig. 41). A prerequisite is the occurrence of brittle fracture along these interfaces. To avoid contamination from the air, the samples are fractured in the UHV chamber of the Auger spectrometer. Numerous studies have shown that embrittlement, stress corrosion cracking and grain growth inhibition are due to an enrichment of impurities in grain boundaries [158]. Intergranular fracture studies in steels have shown S, Sb, P, Sn and Ni to segregate at grain boundaries [159—161]. In tungsten, the amount of grain boundary segregation of P was shown to correlate with the increase of the ductile—brittle transition temperature [162]. Grain boundary segregation of S and Sn in Fe [163] and of Bi in Cu [164] have been thoroughly studied by AES.

AES fracture surface studies in sintered ceramic materials [79,97,165—167] have shown the decisive role of sintering additives such as MgO, which tend to reduce the mechanical strength and enhance high temperature creep [79]. An example of the differences in composition observed in an AES intergranular fracture study of hot-pressed silicon nitride with Al$_2$O$_3$ or MgO as additives is shown in Fig. 43 [97]. In the depth profile of the ratio of the O and N Auger peak-to-peak heights for both materials, little oxygen enrichment is observed in the case of Al$_2$O$_3$, but there is a pronounced enrichment of oxygen extending to about 30 Å into the bulk in the case of MgO additive. This is presumably due to a glassy intergranular

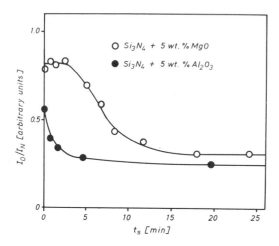

Fig. 43. Ratio of oxygen to nitrogen APPH, I_0/I_N, versus sputtering time, t_s, for silicon nitride hot pressed with MgO (○) and with Al_2O_3 (●) [97].

phase leading to reduced mechanical strength and high temperature creep resistance [79,167]. Table 4 shows a comparison of the composition derived from eqn. (18) before (a) and after (b) sputter removal of the fracture surface with the composition expected from the powder metallurgical processing (c) [97].

TABLE 4

Comparison of two silicon nitride samples (sample 1: Si_3N_4 + 5 wt.% MgO; sample 2: Si_3N_4 + 5 wt.% Al_2O_3) determined by AES at fracture surface using eqn. (18)

(a) Composition (at.%) after fracture; (b) after sputtering; (c) composition data from the producer [97].

Element	Sample 1			Sample 2		
	(a)	(b)	(c)	(a)	(b)	(c)
Si	26	46	40.7	25	37	40
N	40	47	54.3	54	53	54.3
O	19	5	2.5	12	3	2.9
Mg	7		2.5			
Al				4	3	2.0

156

AES has found a widespread use in fundamental surface science. Monitoring the alterations of surface composition (under UHV conditions) which occur during "in situ" thermal treatment has led to a better understanding of such phenomena as equilibria and kinetics of surface segregation, surface diffusion and the effect of ion bombardment on surface composition of alloys. Furthermore, the interaction of surfaces with controlled gas atmospheres, as in adsorption, desorption, oxidation and catalytic activity studies has been investigated.

Equilibrium surface segregation has been extensively studied in binary alloys (Cu—Al [168], Cu—Ni [169,225], Cu—Au [170,171], Ag—Au [172], Ni—Au [173], Pt—Sn [174]) and in dilute solutions [69,98,99,175—179]. Surface enrichment of the species with the lower surface energy is generally observed. A typical result of the surface coverage of the segregant versus the sample temperature is shown in Fig. 44 for oxygen dissolved in niobium [69]. The thermodynamics of the observed surface segregation can be expressed by a Langmuir—McLean equilibrium equation which allows the determination of the segregation enthalpy (about 60 kJ/mole for O in Nb) [69,176]. Observation of the time dependence of the surface enrich-

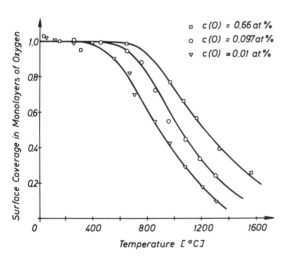

Fig. 44. Equilibrium surface segregation of oxygen in niobium. At temperature measurements of oxygen Auger peak-to-peak heights calibrated in fractions of a monolayer; $c(O)$ = bulk concentration of oxygen [69].

References pp. 166—172
157

ment (after sputtering cleaning) at different sample temperatures allows the determination of the bulk diffusion coefficient of the segregating element to be made as shown for P [180] and for C and S [237] in Fe and for Sn in Cu [98]. Depth profiling confirms the restriction of the enrichment to the outermost surface layers [69,98].

Surface diffusion can be studied by monitoring the spreading of an element on a surface, preferably by high spatial resolution AES. This was demonstrated for Sn on Au [181] and for Pd on W (110) surfaces [182].

Adsorption and desorption are readily observed by means of AES. Many metal—oxygen [183—186] and metal—nitrogen [187,188] adsorption systems have been investigated. Oxidation studies of Ni single crystals have given evidence for the chemisorption of oxygen, its penetration into the selvedge and the growth of bulk oxides [189]. Stimulated desorption by the primary electron beam can be directly studied [135], but it may be a source of error in adsorption experiments [258].

The reaction kinetics of oxygen with surface impurities such as sulphur have been studied at Ni [190] and Cu [191] surfaces by observation of the decrease of the S Auger peak and the increase of the oxygen signal in conjunction with mass spectroscopic gas atmosphere control.

12. Recent developments

(A) HIGH LATERAL RESOLUTION AES

In recent years, there has been a continuous effort to improve the inherent capability of micro-local analysis in AES. Whereas depth resolution is physically limited by the escape depth of Auger electrons, the lateral resolution depends primarily on the diameter of the electron beam. By rastering the electron beam and monitoring the Auger signal, scanning Auger images of a surface are obtained [231]. A first step to localized analysis was done with the introduction of the CMA with an electrostatic-focussed electron gun for typical beam diameters of 50 μm. Further improvement led to commercial systems of 3 μm and 0.5 μm lateral resolution. Even better focussing is obtained by electromagnetic lenses as applied in scanning electron microprobes (SEM). A combination of the CMA analyzer with commercial SEM's has proved useful for high resolution AES [10,192,193]. Due to the limited brightness of hot filament cathodes, beam currents of $\leqslant 10^{-8}$ A are obtained. Therefore, the noise figure [cf. eqn. (5)] requires long sampling times for sensitive AES detection which enhances contamination problems from the residual gas pressure in normal SEM systems [192]. To meet this problem, SEM devices with UHV capability have been developed with reported lateral resolution $\leqslant 0.3\,\mu$m [194,195].

Recently developed instruments with cold cathode emission are especially suited for high lateral resolution AES because of their increased beam brightness [50]. An ultimate resolution of 30 nm has been reported [50]. A further increase in resolution seems to be prevented by electron scattering effects which tend to screen the Auger analysis spot [245].

A general problem in high lateral resolution AES is electron beam effects (see Sect. 10.C) due to the high current density ($>$10 A/cm^2). The usefulness of the Auger microprobe in practical problems has been demonstrated for semiconductors and metals. Various examples

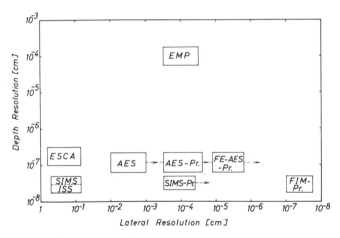

Fig. 45. Comparison of spatial resolution performance of some typical surface analysis techniques. ESCA = electron spectroscopy for chemical analysis; AES, AES-Pr. = high resolution AES Probe; FE-AES Pr. = AES high resolution with field emission cathode; SIMS = secondary ion mass spectrometry; SIMS-Pr. = SIMS-Probe, ion microprobe mass analysis; FIM-Pr. = atom probe in field ion microscopy; EMP (electron microprobe) is shown for comparison as a method not directly applicable to surface analysis in the nanometer region.

may be found in refs. 17 and 50. Compared with other instrumental microanalysis techniques (Fig. 45), its ultimate lateral resolution is only surpassed by the atom probe in field ion microscopy (FIM-Pr).

(B) HIGH ENERGY RESOLUTION AES

In quantitative AES, the energy resolution of the Auger electron analyzer should be matched to the natural line width of the most prominent Auger peaks. Therefore, an energy resolution of about 0.5% is appropriate. However, high energy resolution is extremely useful for obtaining chemical bonding information from AES which is revealed in the fine structure (see Sect. 7.C). High resolution with the CMA can be obtained by pre-retardation of the electrons before they enter the analyzer [196] with the aid of a hemispherical grid in front of the analyzer entrance. Because the absolute CMA energy resolution is proportional to the pass energy, pre-retardation improves the resolution substantially [197]. In practice, the obtainable resolution is limited by the CMA transmission (which is also proportional to the pass energy, see Sect. 5.B) and by the need for suppression of any stray magnetic fields in low energy electron spectroscopy. Commercially available CMA with pre-retarding facility (double pass analyzer) [198], suitable also for ESCA work and spherical deflection analyzers, are frequently used in high energy resolution AES [224].

(C) DATA ACQUISITION AND PROCESSING

The normally recorded $\mathrm{d}N(E)/\mathrm{d}E$ spectra are of limited value for quantitative AES because their line shape and, hence, the peak-to-peak amplitude is dependent on instrumental parameters (see Sect. 5.B) and on various sample parameters (see Sect. 7.C). Therefore, a direct evaluation of the area under the Auger peak is highly desirable. After subtraction of the background, integration of the $N(E)$ distribution gives a measure of the Auger current.

Background subtraction can be made by simple linear interpolation between the background-signal transition points or by the use of fitting polynomes [200]. Due to the monotonously varying background, multiple differentiation and subsequent re-integration (dynamical background subtraction) [201,202] can be successfully applied. Background subtraction and integration techniques are facilitated by the recent development in automation of AES [199]. Here, the analog method of detection with a lock-in amplifier is replaced by a digital counting technique using a voltage-to-frequency converter at the multiplier output. Thus an $EN(E)$ spectrum is obtained with a cylindrical mirror analyzer. The digitized signals can be stored and further processed by an integrated computer; they can be smoothed, integrated, differentiated, etc. [227,236].

The data can also be used to give elemental surface maps in scanning Auger microscopy as well as depth profiles at several points of the sample during a single sputter profiling run.

It should be pointed out that data processing for quantification is not free from arbitraries such as the selection of the final background point or the cut-off point at the low energy tail. Though pitfalls in quantitative evaluation are less probable compared with the differentiated spectrum technique, the accuracy still remains limited by all the physical features on which it relies (see Sects. 8 and 9). It is felt that the ease of quantitative data evaluation by automatic correction with the analyzer parameters and those of Auger transitions [199] will promote their use in the future.

13. Comparable analytical techniques

The most common surface analytical techniques may be subdivided into electron, photon and ion spectroscopies. Either the escape depth of the detected particles or the excitation depth should be of the order of a few atomic layers to establish their surface analysis ability [248]. A comparison of some of the more customary techniques in terms of their lateral resolution and depth resolution is shown in Fig. 45. Figure 46 shows a comparison with respect to their analysis volume.

(A) ELECTRON SPECTROSCOPIES

Among those most directly related to AES are other electron spectroscopic techniques like disappearance potential spectroscopy (DAPS) [18] or ionization loss spectroscopy (ILS) [18,203], and X-ray photoelectron spectroscopy (ESCA or XPS, UPS) [43,206,230].

Energy loss peaks are frequently encountered in AES due to plasma or interband transition excitations (see Sect. 4). Plasmon

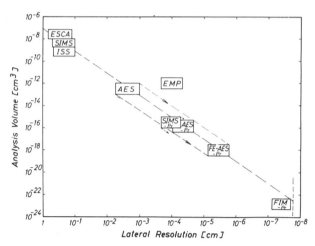

Fig. 46. Comparison of analysis volume versus lateral resolution of the methods compiled in Fig. 45.

peaks contribute to the low energy side of the Auger peaks. Ionization loss peaks are generated by primary electrons with energy E_p above the ionized core level E_W. Therefore, they appear at a detected energy $E_p - E_W$ and can be used for the identification of elements similar to ESCA. In contrast to Auger lines, loss peaks move together with varying primary energy E_p (see Sect. 4) and are therefore measured at fixed analyzer energy and programmed E_p [18]. Their high signal-to-noise ratio due to the large background from multiple inelastic scattering events limits their usefulness in general surface analysis. In fundamental research, the narrow line width and simplicity of ILS spectra and their sensitivity to chemical effects [203,204] seem to be very promising.

Monitoring the "disappearance" of electrons from the elastic peak if the primary energy is tuned over an ionization threshold is the principle of disappearance potential spectroscopy [18,247] a technique complementary to ILS.

In ESCA (XPS) the electron spectroscopy technique is quite similar to (high energy resolution) AES except that monoenergetic X-ray photons are used for excitation [205,206,230,233]. The energy of the generated photoelectrons is a measure of the binding energy. Since only one energy level is involved, chemical shifts can be directly studied and are the main field of application of ESCA. The

162

restrictions in generating high intensity and focussed X-ray beams pose certain disadvantages (low spatial resolution in the mm range and low sensitivity requiring higher sampling times) compared with AES. However, they are outweighed by the fact that X-rays are much less destructive than electron beams (see Sect. 10.C), which is of special importance in the study of organic materials or adsorbates [206].

(B) PHOTON SPECTROSCOPIES

Optical spectroscopy can be applied in surface analysis if the emission is restricted to surface constituents. Recently developed techniques like SCANIIR (surface composition by analysis of neutral and ion impact radiation), IBSCA (ion beam spectrochemical analysis) or GDOS (glow discharge optical spectroscopy) show that the optical radiation observed during ion bombardment of solids, which is due to radiation from excited states of sputtered surface constituents, can be used in surface and thin film analysis [213,214]. A special advantage lies in the study of insulating materials (e.g. glasses [213]) where radiationless transitions are suppressed and detection sensitivities in the ppm range have been obtained [214]. Due to the same ion beam used for probing, the technique should be easily combined with SIMS with the advantage of complementary information.

Another method based on the detection of electromagnetic radiation is appearance potential spectroscopy of soft X-rays (SXAPS) [204]. In this technique, only the amount of emitted photons due to electron excitation is detected [215,216]. If the current collected by a photocathode is monitored during continuous increase of the primary electron energy E_p its stepwise increase at energy E_p which corresponds to a discrete binding energy allows the determination of core electron levels. SXAPS has been frequently used for the investigation of chemical effects [217,218]. A large background (due to Bremsstrahlung) limits the application in surface analysis.

(C) ION SPECTROSCOPIES

These techniques comprise ion scattering spectroscopy (ISS), secondary ion mass spectroscopy (SIMS), sputtered neutrals mass spectroscopy or glow discharge mass spectroscopy (SNMS, GDMS), and atom probe field ion microscopy (AP-FIM). Mass analysis of ions emitted due to other excitation processes, e.g. electron-induced or -stimulated desorption (EID, ESD) can also be applied as an analytical tool [224].

Within the more commonly used surface analysis techniques, ISS with low energy ions is most surface specific. A monoenergetic beam of ions (0.1–10 keV) strikes the surface and the energy distribution of ions scattered in a particular angle is measured [207]. The energy spectrum contains information about the mass and number of the atoms in the very first layer of a solid. For the heavier elements, the sensitivity is comparable with AES. ISS is highly sensitive to surface structure, whereas chemical effects are not revealed. Its use as a surface analytical tool is only in its infancy, but the efforts in its development [208] in recent years have resulted in a commercial instrument becoming available (3M Company, U.S.A.).

The related method of Rutherford back-scattering spectroscopy (RBS) with high energy primary ions in the 100 keV range is less suited to surface studies but a powerful quantitative tool for non-destructive thin film analysis with depth resolution of about 10 nm [243].

SIMS is presumably an analytical technique with the broadest capability for both surface and bulk analysis of solids [209,210,234]. Incident high energy ions cause the emission of particles from the sample surface (neutrals and ions). The most outstanding features compared with AES are its high detection sensitivity (typical in the ppm range for most elements) and its ability to detect H and He, as well as isotopes. Indirect chemical information is obtained from molecule ions. Spatial resolution in the ion microprobe technique ($\leqslant 1 \mu$m) is comparable with AES microprobes. The main disadvantages are matrix effects resulting in composition-dependent sensitivity alterations for the elements. Sputtering with reactive ions has proved useful for quantitative work [113,121,210]. In this respect, post-ionization of the material removed by sputtering [211] as in glow discharge mass spectrometry (GDMS) largely avoids concentration and chemical effects on quantitative analysis but is generally restricted to low lateral resolution and is highly prone to contamination from the discharge chamber walls [67,119].

Since sputter erosion of the surface destroys the original surface, SIMS (and GDMS) inherently shows a time dependence which contains information about in-depth distribution of elemental con-

TABLE 5

Comparison of the characteristic features of Auger electron spectroscopy (AES), Photoelectron Spectroscopy (ESCA), Secondary Ion Mass Spectrometry (SIMS) and Ion Scattering Spectroscopy (ISS).

	AES	ESCA	SIMS	ISS
Principal excitation	e^-	$h\nu$	$I^\pm_{prim}$	$I^\pm_{prim.}$
Principal emission	e^- (E)	e^- (E)	$I^\pm_{sec}$ (m/e)	$I^\pm_{prim.}$ (E)
Information depth (atomic layers)	2—10	2—10	1—3	1
Detection sensitivity				
ppm	1000	1000	1	1000
g/cm^2	10^{-10}	10^{-10}	10^{-13}	10^{-10}
Sensitivity differences of elements (factor)	10	10	10^5	10^2
Detection of				
elements	$Z > 2$	$Z > 1$	All	$Z > 1$
isotopes	No	No	Yes	Yes
chemical bond	Special cases	Yes	Yes	No
depth profiles	+ sputtering	+ sputtering	Yes	Yes

centration which has led to its widespread use in thin film analysis [67].

The ultimate possibility in microanalysis seems to be the atom probe field ion microscopy (AP-FIM) [212], where single atoms can be detected. However, due to its sophisticated experimental and theoretical implications, it is generally used in fundamental research and will not be looked upon as a mature surface analysis tool in this context.

The basic characteristics of the most frequently used surface analysis techniques ESCA, AES, SIMS and ISS are compared in Table 5.

Acknowledgements

The author would like to thank Dr. J. Erlewein for useful comments and Miss U. Nagorny for skilful preparation of the manuscript.

References

1 P. Auger, Compt. Rend., 177 (1923) 169.
2 P. Auger, J. Phys. Radium, 6 (1925) 205.
3 P. Auger, Ann. Phys. (Paris), 6 (1926) 183.
4 P. Auger, Surf. Sci., 48 (1975) 1.
5 J.J. Lander, Phys. Rev., 91 (1953) 1382.
6 L.A. Harris, J. Appl. Phys., 39 (1968) 1419.
7 L.A. Harris, J. Appl. Phys., 39 (1968) 1428.
8 R.E. Weber and W.T. Peria, J. Appl. Phys., 38 (1967) 4355.
9 P.W. Palmberg, G.K. Bohn and J.C. Tracy, Appl. Phys. Lett., 15 (1969) 254.
10 N. MacDonald, Appl. Phys. Lett., 16 (1970) 76.
11 P.W. Palmberg and H.L. Marcus, Trans. Am. Soc. Met., 62 (1969) 1016.
12 D.T. Hawkins, Bibliography of Auger Electron Spectroscopy, IFI, Plenum Press, New York, 1977.
13 N.J. Taylor, in R.F. Bunshah (Ed.), Techniques of Metals Research, Vol. 7, Interscience, New York, 1971, p. 28.
14 E. Bauer, Z. Metallkd., 63 (1972) 437.
15 J.C. Rivière, Contemp. Phys., 19 (1973) 513.
16 C.C. Chang, in R.F. Kane and G.B. Larabee (Eds.), Characterization of Solid Surface, Plenum Press, New York, 1974, p. 509.
17 A. Joshi, L.E. Davis and P.W. Palmberg, in A.W. Czanderna (Ed.), Methods of Surface Analysis, Elsevier, Amsterdam, 1975, p. 159.
18 J. Kirschner, in H. Ibach (Ed.), Electron Spectroscopy for Surface Analysis, Springer Verlag, Heidelberg, 1977, p. 59.
19 M.P. Seah, Surf. Sci., 32 (1972) 703.
20 P.W. Palmberg, Anal. Chem., 45 (1973) 549A.
21 C.J. Powell, Surf. Sci., 49 (1974) 29.
22 H.D. Hagstrum, J.E. Rowe and J.C. Tracy, in R.B. Anderson and P.T. Dawson (Eds.), Experimental Methods in Catalytic Research, Vol. III, Academic Press, New York, 1976, p. 42.
23 P.W. Palmberg, J. Vac. Sci. Technol., 9 (1972) 160.
24 E.H.B. Burhop, The Auger Effect and other Radiationless Transitions, Cambridge University Press, London, 1952.
25 K.D. Sevier, Low Energy Electron Spectrometry, Wiley Interscience, New York, 1972.
26 D. Chattarji, The Theory of Auger Transitions, Academic Press, London, 1976.
27 M.F. Chung and L.H. Jenkins, Surf. Sci., 22 (1970) 479.
28 J.A. Bearden and A.F. Burr, Rev. Mod. Phys., 39 (1967) 125.
29 D. Coster and R.L. Kronig, Physica, 2 (1935) 13.
30 E.J. McGuire, in B. Crasemann (Ed.), Atomic and Inner Shell Processes, Vol. 1, Academic Press, London, 1975, p. 293.
31 J.P. Coad, Z. Phys., 244 (1971) 19.
32 H.E. Bishop and J.C. Rivière, J. Appl. Phys., 40 (1969) 1740.
33 L.E. Davis, N.C. MacDonald, P.W. Palmberg, G.E. Riach and R.E. Weber,

Handbook of Auger Electron Spectroscopy, Physical Electronics Ind.,
Eden Prairie, 1976.

34 W.A. Coghlan and R.E. Clausing, USAEC Rep. ORNL-TM-3576, U.S. Dept.
of Commerce, Springfield, 1971.

35 E. Bauer, Vacuum, 22 (1972) 539.

36 M. Gryzinsky, Phys. Rev. A, 138 (1965) 336.

37 D.M. Smith and T.E. Gallon, J. Phys. D., 7 (1974) 151.

38 J.J. Vrakking and F. Meyer, Surf. Sci., 47 (1975) 50.

39 H.W. Drawin, Z. Phys., 164 (1961) 513.

40 K. Goto and K. Ishikawa, Surf. Sci., 47 (1975) 475.

41 J.P. Coad, H.E. Bishop and J.C. Rivière, Surf. Sci., 21 (1970) 253.

42 R.G. Steinhardt and E.H. Serfass, Anal. Chem., 23 (1951) 1585.

43 K. Siegbahn, C. Nordling, A. Fahlmann, R. Nordberg, K. Hamerin, J. Hed-
mann, G. Johansson, T. Bergmark, S.-E. Karlsson, T. Lindgren and B.
Lindberg, ESCA: Atomic, Molecular and Solid State Structure Studied by
Means of Electron Spectroscopy, Almquist and Wiksells, Uppsala, 1967.

44 R.G. Musket and W. Bauer, Appl. Phys. Lett., 20 (1972) 422.

45 K.O. Groeneveld, R. Mann, W. Meckbach and R. Spohr, Vacuum, 25
(1975) 9.

46 F. Louchet, L. Viel, C. Benazeth, B. Fagot and N. Colombie, Radiat. Eff.,
14 (1972) 123.

47 J. Kempf and G. Kaus, Appl. Phys., 13 (1977) 261.

48 G.N. Ogustson, Rev. Mod. Phys., 44 (1972) 1.

49 N.C. MacDonald, G.E. Riach and R.L. Gerlach, Res. Dev., 27 (8) (1976)
42.

50 J.A. Venables, A.P. Janssen, C.J. Harland and B.A. Joyce, Phil. Mag., 34
(1976) 495.

51 N.J. Taylor, Rev. Sci. Instrum., 40 (1969) 792.

52 P.A. Readhead, J.P. Hobson and F.V. Kornelsen, The Physical Basis of
Ultra High Vacuum, Chapman and Hall, London, 1968.

53 J.A. Simpson, Rev. Sci. Instrum., 32 (1961) 1283.

54 E. Blauth, Z. Phys., 147 (1957) 228.

55 H. Sar-El, Rev. Sci. Instrum., 38 (1967) 1210.

56 H. Sar-El, Rev. Sci. Instrum., 39 (1968) 533.

57 N.J. Taylor, J. Vac. Sci. Technol., 6 (1969) 241.

58 P.W. Palmberg and T.N. Rhodin, J. Appl. Phys., 39 (1968) 2425.

59 G.A. Harrower, Rev. Sci. Instrum., 26 (1955) 850.

60 A.L. Hughes and V. Rojansky, Phys. Rev., 34 (1929) 284.

61 H. Hafner, J.A. Simpson and C.E. Kuyatt, Rev. Sci. Instrum., 39 (1968) 33.

62 W. Färber and P. Braun, Vak. Tech., 8 (1974) 239.

63 P. Braun and W. Färber, Surf. Sci., 47 (1975) 57.

64 G. Betz, P. Braun and W. Färber, J. Appl. Phys., 48 (1977) 1404.

65 E. Bauer, H. Poppa and F. Bonczek, J. Appl. Phys., 45 (1974) 5164.

66 M.P. Seah, Surface Sci., 40 (1973) 595.

67 J.W. Coburn and E. Kay, Crit. Rev. Solid State Sci., 4 (1974) 561.

68 A. Zalar, A. Banovec and S. Hofmann, Proc. 7th Int. Vac. Congr. and 3rd
Int. Conf. Solid Surfaces, Vienna, 1977, Vol. III, p. 2303.

69 S. Hofmann, G. Blank and H. Schultz, Z. Metallkd., 67 (1976) 189.

70 E.N. Sickafus and D.M. Holloway, Surf. Sci., 51 (1975) 139.

71 W.H.R. Losch, J. Vac. Sci. Technol., 13 (1976) 737.
72 K. Stadler, Vacuum, 22 (1972) 553.
73 B.D. Powell and D.P. Woodruff, Surf. Sci., 33 (1972) 437.
74 G. Allie, E. Blanc, D. Dufayard and P. Hayman, Surf. Sci., 47 (1975) 635.
75 J. Erlewein, Thesis, University of Stuttgart, 1977.
76 T.W. Haas and J.T. Grant, Surf. Sci., 24 (1971) 332.
77 T.W. Haas, J.T. Grant and G.J. Dooley, J. Appl. Phys., 43 (1972) 1853.
78 R. Heckingbottom and P.R. Wood, Surf. Sci., 36 (1973) 594.
79 S. Hofmann and L.J. Gauckler, Powder Metall. Int., 6 (1974) 90.
80 P.H. Holloway, Surf. Sci., 54 (1976) 506.
81 D.T. Quinto and W.D. Robertson, Surf. Sci., 27 (1971) 645.
82 S. Hofmann, Mikrochim. Acta Suppl., 7 (1977) 109.
83 K. Müller, in Springer Tracts in Modern Physics, Vol. 77, Springer, Heidel-
    berg—Berlin—New York, 1975, pp. 97—125.
84 R. Weissmann, R. Koschatzky, W. Schnellhammer and K. Müller, Appl.
    Phys., 13 (1977) 43.
85 W.M. Mularie and W.T. Peria, Surf. Sci., 26 (1971) 125.
86 F.N. Sickafus, J. Vac. Sci. Technol., 11 (1974) 308.
87 P. Braun, G. Betz and W. Färber, Mikrochim. Acta Suppl., 5 (1974) 365.
88 S. Ferrer and J.M. Rojo, Solid State Commun., 24 (1977) 339.
89 R.E. Weber and A.L. Johnson, J. Appl. Phys., 40 (1969) 314.
90 P.W. Palmberg, J. Vac. Sci. Technol., 13 (1976) 214.
91 L.E. Davis and A. Joshi, in R.S. Carbonara and J.R. Cuthill (Eds.), Surface
    Analysis Techniques for Metallurgical Applications, American Society for
    Testing Materials, 1976, ASTM STP 596, p. 52.
92 S. Hofmann and G. Blank, Forschungsberichte aus der Wehrtechnik, Nr. 20,
    Bonn, 1976, p. 219.
93 T.E. Gallon, Surf. Sci., 17 (1969) 486.
94 C. Argile and G.F. Rhead, Surf. Sci., 53 (1975) 659.
95 G. Rhead, J. Vac. Sci. Technol., 13 (1976) 603.
96 P.W. Palmberg, Appl. Phys. Lett., 13 (1968) 183.
97 S. Hofmann, L.J. Gauckler and L. Tillmann, Mikrochim. Acta Suppl.,
    6 (1975) 373.
98 S. Hofmann and J. Erlewein, Scr. Metall., 10 (1976) 857.
99 S. Hofmann and J. Erlewein, Surf. Sci., 68 (1977) 71.
100 G.K. Wehner, in A.W. Czanderna (Ed.), Methods of Surface Analysis,
    Elsevier, Amsterdam, 1975, p. 5.
101 S. Hofmann, Appl. Phys., 9 (1976) 59.
102 G. Carter, J.S. Colligon, Ion Bombardment of Solids, Heinemann, London,
    1968.
103 R.E. Honig, Thin Solid Films, 31 (1976) 89.
104 S. Hofmann, Appl. Phys., 13 (1977) 205.
105 D.W. Coburn, J. Vac. Sci. Technol., 13 (1976) 1037.
106 H. Shimzu, M. Ono and K. Nakayama, Surf. Sci., 36 (1973) 817.
107 H.J. Mathieu and D. Landolt, Surf. Sci., 53 (1975) 228.
108 P.S. Ho, J.E. Lewis, H.S. Wildmann and J.K. Howard, Surf. Sci., 57 (1976)
    393.
109 J.A. McHugh, Radiat. Eff., 21 (1974) 209.

110 H.J. Mathieu, D.E. McClure and D. Landolt, Thin Solid Films, 38 (1976) 281.
111 H.F. Winters and J.W. Coburn, Appl. Phys. Lett., 28 (1976) 176.
112 C.G. Pantano, D.B. Dove and G.Y. Onoda, J. Vac. Sci. Technol., 13 (1976) 414.
113 W.O. Hofer and H. Liebl, J. Appl. Phys., 8 (1975) 359.
114 S. Hofmann, J. Erlewein and A. Zalar, Thin Solid Films, 43 (1977) 275.
115 A. Benninghoven, Z. Phys., 230 (1970) 403.
116 P.S. Ho and J.E. Lewis, Surf. Sci., 55 (1976) 335.
117 S. Hofmann, Proc. 7th Int. Vac. Congr. and 3rd Int. Conf. Solid Surfaces, Vienna, 1977, Vol. 3, p. 2613.
118 P.H. Holloway, J. Electron Spectrosc. Relat. Phenom., 7 (1975) 215.
119 J.W. Coburn, F.W. Eckstein and E. Kay, J. Appl. Phys., 46 (1975) 2828.
120 H.W. Werner, Acta Electron., 19 (1976) 53.
121 M. Bernheim and G. Slodzian, Surf. Sci., 40 (1973) 169.
122 B.R. Martin, Statistics for Physicists, Academic Press, London, 1971.
123 J.A. McHugh, in K.F. Heinrich and D.E. Newbury (Eds.) Workshop on SIMS, Gaithersburgh, 1975, NBS Spec. Publ. 427, p. 79.
124 S. Hofmann and A. Zalar, Thin Solid Films, 39 (1976) 219.
125 S. Hofmann, Z. Anal. Chem., 285 (1977) 177.
126 L.L. Chang and A. Koma, Appl. Phys. Lett., 29 (1976) 128.
127 C.R. Helms, C.M. Garner, J. Müller, I. Lindau, S. Schwarz and W.E. Spicer, Proc. 7th Int. Vac. Congr. and 3rd Int. Conf. Solid Surfaces, Vienna, 1977, Vol. 3, p. 2241.
128 L.B. Castle and L. Hazell, J. Electron. Spectrosc. Relat. Phenom., 12 (1977) 195.
129 G. Tölg, Mikrochim. Acta Suppl., 7 (1977) 1.
130 L. McDonell, D.P. Woodruff and B.W. Holland, Surf. Sci., 51 (1975) 249.
131 J.R. Noonan, D.M. Zehner and L.H. Jenkins, J. Vac. Sci. Technol., 13 (1976) 183.
132 J.P. Coad, M. Gettings and J.C. Rivière, Faraday Discuss. Chem. Soc., 60 (1976) 269.
133 S. Thomas, J. Appl. Phys., 45 (1974) 161.
134 T.E. Madey and J.T. Yates, J. Vac. Sci. Technol., 8 (1971) 525.
135 R.G. Musket and J. Ferrante, Surf. Sci., 21 (1970) 440.
136 S. Thomas, Appl. Phys. Lett., 24 (1974) 1.
137 J. Schlichting and S. Hofmann, High Temp. High Pressures, to be published.
138 R. Riwan, Surf. Sci., 27 (1971) 267.
139 J. Ahn, C.R. Perleberg, D.L. Wilcox, J.W. Coburn and H.F. Winters, J. Appl. Phys., 46 (1975) 4581.
140 P. Braun, G. Betz and W. Färber, Mikrochim. Acta Suppl., 7 (1977) 129.
141 P.F. Kane and C.B. Larabee, Anal. Chem., 49 (1977) 221R.
142 R.E. Thomas and G.A. Haas, J. Appl. Phys., 46 (1975) 963.
143 K. Beck and K.L. Schiff, Z. Werkstofftech., 8 (1975) 117.
144 M. Murko-Jezovsek, F. Bercelj and B. Jenko, Proc. 7th Int. Vac. Congr. and 3rd Int. Conf. Solid Surfaces, Vienna, 1977, Vol. 3, p. 2343.
145 J.J. Uebbing, J. Appl. Phys., 41 (1970) 802.
146 N.C. MacDonald and G.F. Riach, Electron. Packag. Prod., 13 (1973) 50.

147 A. Zalar, A. Banovec and S. Hofmann, Proc. 7th Int. Vac. Congr. and 3rd Int. Conf. Solid Surfaces, Vienna, 1977, Vol. 3, p. 2303.
148 D.H. Buckley and S.V. Pepper, Am. Soc. Lubr. Eng. Trans., 15 (1972) 252.
149 N.A. Gjostein and N.G. Chauka, J. Test. Eval., 1 (1973) 183.
150 J.B. Lumsden and R.W. Staehle, in R.S. Carbonara and J.R. Cuthill (Eds.), Surface Analysis Techniques for Metallurgical Applications, ASTM 1976, ASTM STP 596, p. 39.
151 W. Hemminger, S. Hofmann and U. Holz, Unfallheilkd. Traumatol., 79 (1976) 423.
152 G. Betz, G.K. Wehner, L.E. Toth and A. Joshi, J. Appl. Phys., 45 (1974) 5312.
153 G.A. Somorjai, Catal. Rev., 7 (1972) 87.
154 M.M. Bhasin, J. Catal., 34 (1974) 356.
155 P.M. Hall and J.M. Morabito, Surf. Sci., 54 (1976) 79.
156 H.S. Wildman, J.K. Howard and P.S. Ho, J. Vac. Sci. Technol., 12 (1975) 75.
157 T.J. Chuang and K. Wandelt, Surf. Sci., Proc. 7th Int. Vac. Congr. and 3rd Int. Conf. Solid Surfaces, Vienna, 1977, Vol. 3, p. 2091.
158 M.P. Seah, Surf. Sci., 53 (1975) 168.
159 H.L. Marcus and P.W. Palmberg, Trans. AIME, 245 (1969) 1164.
160 R. Viswanathan, Metall. Trans., 2 (1971) 809.
161 R. Viswanathan and T.P. Sherlock, Metall. Trans., 3 (1972) 459.
162 A. Joshi and D.F. Stein, Metall. Trans., 1 (1970) 2543.
163 M.P. Seah and E.D. Hondros, Proc. R. Soc. London, Ser. A, 335 (1973) 191.
164 B.D. Powell and D.P. Woodruff, Phil. Mag., 34 (1976) 169.
165 B.D. Powell and P. Drew, J. Mater. Sci., 9 (1974) 1867.
166 W.C. Johnson, D.F. Stein and R.W. Rice, J. Am. Ceram. Soc., 57 (1974) 342.
167 W.C. Johnson, Metall. Trans., 8A (1977) 1413.
168 J. Ferrante, Acta Metall., 19 (1971) 743.
169 G. Ertl and J. Küppers, J. Vac. Sci. Technol., 9 (1972) 829.
170 J.M. McDavid and S.C. Fain, Surf. Sci., 52 (1975) 169.
171 R.A. Van Santen, L.H. Toneman and R. Bouwman, Surf. Sci., 47 (1975) 64.
172 S.H. Overbury and G.A. Somorjai, Surf. Sci., 55 (1976) 209.
173 J.J. Burton, C.R. Helms and R.S. Polizotti, J. Vac. Sci. Technol., 13 (1976) 204.
174 R. Bouwman, L.H. Toneman and A.A. Holscher, Surf. Sci., 35 (1973) 8.
175 W.P. Ellis, J. Vac. Sci. Technol., 9 (1972) 1027.
176 A. Joshi and M. Strongin, Scr. Metall., 8 (1974) 413.
177 L.C. Isett and M. Blakely, J. Vac. Sci. Technol., 12 (1975) 237.
178 H.J. Grabke, G. Tauber and H. Viefhaus, Scr. Metall., 9 (1975) 1181.
179 M.P. Seah and C. Lea, Phil. Mag., 31 (1975) 627.
180 C.A. Shell and J.C. Rivière, Surf. Sci., 40 (1973) 149.
181 H.G. Tompkins, J. Vac. Sci. Technol., 12 (1975) 650.
182 R. Butz and H. Wagner, Verh. Dtsch. Phys. Ges., 1 (1977) 353.
183 J.C. Tracy and J.M. Blakely, Surf. Sci., 15 (1969) 257.

184 H.H. Farrell, H.S. Isaacs and M. Strongin, Surf. Sci., 38 (1973) 31.
185 E.B. Bas and U. Bänninger, Surf. Sci., 49 (1974) 1.
186 N. Pacia, J.A. Dumesic, B. Weber and A. Cassuto, Faraday Trans. Chem. Soc., 72 (1976) 1919.
187 J.M. Dickey, Surf. Sci., 50 (1975) 515.
188 G. Ertl, M. Grunze and M. Weiss, J. Vac. Sci. Technol., 13 (1976) 314.
189 P.H. Holloway and J.B. Hudson, Surf. Sci., 43 (1974) 141.
190 P.H. Holloway and J.B. Hudson, Surf. Sci., 25 (1972) 56.
191 H.P. Bonzel, Surf. Sci., 27 (1971) 387.
192 R. Holm and S. Storp, Mikrochim. Acta Suppl., 7 (1977) 171.
193 R. Le Bihan, Proc. 7th Int. Vac. Congr. and 3rd Int. Conf. Solid Surfaces, Vienna, 1977, Vol. 3, p. 2351.
194 T. Ishida, M. Uchiyama, Z. Oda and H. Hashimoto, J. Vac. Sci. Technol., 13 (1976) 711.
195 B.D. Powell, D.P. Woodruff and B.W. Griffiths, J. Phys. E, 8 (1975) 548.
196 L.R. Gerlach, J. Vac. Sci. Technol., 10 (1973) 122.
197 P.W. Palmberg, J. Electron Spectrosc. Relat. Phenom., 5 (1974) 691.
198 P.W. Palmberg and W.M. Riggs, Proc. 7th Int. Vac. Congr. and 3rd Int. Conf. Solid Surfaces, Vienna, 1977, Vol. 3, p. 2617.
199 N.J. Taylor, M.E. Strausser and T.A. Pandolfi, Proc. 7th Int. Vac. Congr. and 3rd Int. Conf. Solid Surfaces, Vienna, 1977, Vol. 3, p. 2621.
200 P. Staib and J. Kirschner, Appl. Phys., 3 (1974) 421.
201 J.E. Houston, Rev. Sci. Instrum., 45 (1974) 897.
202 J.T. Grant, M.P. Hooker and T.W. Haas, Surf. Sci., 51 (1975) 318.
203 R.L. Gerlach, in D.A. Shirley (Ed.), Electron Spectroscopy, North-Holland, Amsterdam, 1972, p. 885.
204 A. Koma and R.R. Ludeke, Phys. Rev. Lett., 35 (1975) 107.
205 R. Holm, Angew. Chem., 83 (1971) 632.
206 W.M. Riggs and M.J. Parker, in W. Czanderna (Ed.), Methods of Surface Analysis, Elsevier, Amsterdam, 1975, p. 103.
207 E. Taglauer and W. Heiland, Appl. Phys., 9 (1976) 261.
208 F.W. Karasek, Res. Dev., 1 (1978) 26.
209 A. Benninghoven, Appl. Phys., 1 (1973) 3.
210 H.W. Werner, Surf. Sci., 47 (1975) 301.
211 H. Oechsner and W. Gerhard, Surf. Sci., 44 (1974) 480.
212 E.W. Müller, in A.W. Czanderna (Ed.), Methods of Surface Analysis, Elsevier, Amsterdam, 1975, p. 329.
213 H. Bach. J. Non-Cryst. Solids, 19 (1975) 65.
214 N.H. Tolk, I.S.T. Tsong and C.W. White, Anal. Chem., 49 (1977) 16A.
215 R.L. Park, J.E. Houston and D.G. Schreiner, Rev. Sci. Instrum., 41 (1970) 1810.
216 R.L. Park, Surf. Sci., 48 (1975) 80.
217 G. Ertl and K. Wandelt, Surf. Sci., 50 (1975) 479.
218 C. Webb and P.M. Williams, Surf. Sci., 53 (1975) 110.
219 P.M. Hall, J.M. Morabito and D.K. Conley, Surf. Sci., 62 (1977) 1.
220 H.J. Mathieu, J.B. Mathieu, D.E. McClure and D. Landolt, J. Vac. Sci. Technol., 14 (1977) 1023.
221 R. Holm and S. Storp, Mikrochim. Acta Suppl., 7 (1977) 139.

222 P. Braun, G. Betz and W. Färber, Mikrochim. Acta Suppl., 7 (1977) 129.
223 C. Le Gressus, D. Massignon, and R. Sopizet, Surf. Sci., 68 (1977) 338.
224 G.C. Allen, P.M. Tucker and R.K. Wild, Surf. Sci., 68 (1977) 469.
225 F.J. Kuijers and V. Ponec, Surf. Sci., 68 (1977) 294.
226 P.M. Hall, J.M. Morabito and J.M. Poate, Thin Solid Films, 33 (1976) 107.
227 J.S. Solomon and W.L. Baun, J. Vac. Sci. Technol., 13 (1975) 375.
228 J. Kirschner and H.W. Etzkorn, Proc. 7th Int. Congr. and 3rd Int. Conf.
    Solid Surfaces, Vienna, 1977, Vol. 3, p. 2213.
229 C.C. Chang, Surf. Sci., 48 (1975) 9.
230 C.R. Brundle, Surf. Sci., 48 (1975) 99.
231 G.W.B. Ashwell, C.J. Todd and R. Heckingbottom, J. Phys. E, 6 (1973)
    435.
232 St.V. Pepper, Rev. Sci. Instrum., 44 (1973) 826.
233 K.L. Cheng and J.W. Prather, Crit. Rev. Anal. Chem., 5 (1975) 37.
234 H. Liebl, J. Phys. E, 8 (1975) 797.
235 H. Schneider, Mikrochim. Acta, II (1977) 437.
236 N.J. Chon, R. Hammer and A. Bednowitz, Rev. Sci. Instrum., 47 (1976)
    559.
237 W.E. Swartz and D.M. Holloway, Appl. Spectrosc., 31 (1977) 210.
238 M.B. Chamberlain and S.C. Lehozky, Thin Solid Films, 45 (1977) 189.
239 B. Carrière and B. Lang, Surf. Sci., 64 (1977) 209.
240 H.W. Werner and N. Warmoltz, Surf. Sci., 57 (1976) 706.
241 N. Saeki and R. Shimizu, Surf. Sci., 71 (1978) 479.
242 R.W. Joyner and J. Rickman, Surf. Sci., 67 (1977) 351.
243 R. Behrisch, B.M. Scherzer and S. Staib, Thin Solid Films, 19 (1973) 57.
244 H.H. Madden, J. Vac. Sci. Technol., 13 (1976) 228.
245 J. Kirschner, Appl. Phys., 14 (1977) 351.
246 C.J. Powell, Appl. Surf. Sci., 1 (1978) 143.
247 J. Kirschner and W. Losch, J. Vac. Sci. Technol., 14 (1977) 1173.
248 C.A. Evans, Jr., Anal. Chem., 47 (1975) 818A.
249 M.P. Seah and W.A. Dench, Nat. Phys. Lab. (U.K.) Div. Chem. Stand. Rep.
    82, 1978.
250 H.E. Bauer, P. Wiedman and H. Seiler, Vak. Tech., 26 (1977) 236.
251 P.M. Hall and J.M. Morabito, Surf. Sci., 67 (1977) 373.
252 C.C. Chang and D.M. Boulin, Surf. Sci., 69 (1977) 385.
253 S. Hofmann and J. Erlewein, to be published in Mikrochim. Acta.
254 D.L. Malm, M.J. Vasile, F.J. Padden, D.B. Dove and C.G. Pantano, J. Vac.
    Sci. Technol., 15 (1978) 35.
255 S. Hofmann and A. Zalar, to be published in Thin Solid Films.
256 H.G. Suzuki and K. Yamamoto, Mater. Sci. Eng., 33 (1978) 57.
257 A.C. Yen, W.R. Graham and G.R. Belton, Metall. Trans., 9A (1978) 31.
258 L. Wiedmann, O. Ganschow and A. Benninghoven, J. Electron Spectrosc.
    Relat. Phenom., 13 (1978) 243.
259 J.H. Onsgaard, P. Morgan and R.P. Creaser, J. Vac. Sci. Technol., 15 (1978)
    44.

Chapter 3

Plasma excitation in spectrochemical analysis

P. TSCHÖPEL

1. Introduction

During the course of the almost explosive developments in science and technology in the last few decades, the demands made on analytical techniques have continually risen. In all this, the requirements of speed and precision have to take second place, since people have always been interested in getting the results of an analysis as quickly as possible, for example in production control of iron and steel works. And the high precision with which analysis could be carried out, even at the beginning of this century, is only seldom required nowadays.

On the other hand, the number of samples, the number of different kinds of sample and the number of elements of interest in a sample is continually increasing, and the concentrations of the elements to be analyzed are reaching lower and lower values. Good powers of detection are necessary, not only for trace analysis in new materials such as semiconductors and reactor materials, but also in, for example, many matrices from environmental research and control, the earth and biological sciences and many others.

In order to solve the given problems optimally, many different principles of analysis, which fulfil the following requirements, must be available.

(1) Power of detection in the low ng/g region or lower, corresponding to an absolute mass of a few ng and less of the element.

(2) Simultaneous or sequential measurement of the elements of interest.

(3) Good reproducibility.

(4) Freedom from systematic errors.

(5) Ease of calibration.

At the moment, however, we are a long way from being able to meet all these trivial requirements at the same time.

Simultaneous or sequential methods are available in, for instance, neutron activation analysis, emission spectroscopy, X-ray fluorescence spectrometry or mass spectrometry. However, with these methods, the excitation of the analytical signal is, in most cases, greatly affected by large scale interference from the matrix and accompanying elements, so that, for very low contents, the result of the analysis can differ from the true value by several orders of magnitude.

Let us look more closely at optical emission spectroscopy (OES) with these considerations in mind.

An excitation source in this principle of analysis has, in the main, three functions to fulfil: (1) to vaporize the sample, (2) to atomize the constituents of the vapour, and (3) to excite as large a proportion as possible of the atoms to emit the analytical signal. These complex procedures are subject to a lot of interference which must be eliminated. In general, the intensity of the emitted analytical line is dependent, not only on the concentration of the element in the sample, but also on the structure of the sample, and partly on its physical nature. The difficulties are encountered chiefly in the vaporization and atomization steps.

The calibration of the procedure, conditional on such interference, is especially difficult in the analysis of solid samples, perhaps by arc or spark excitation, since standard materials are required for this which have exactly the same composition and physical nature as the sample.

The production of such exactly analyzed standards is only worthwhile if there are a large number of similar samples to be investigated by this method, as is the case in, for instance, metal works.

The problem of calibration is somewhat easier to deal with if the sample is isoform, for example, by fusing it with borax. A kind of homogeneity is achieved, eliminating matrix effects. Standard samples in this form are easily produced. But in this way a new source of error appears in the losses due to the high volatility of many elements and compounds.

The aforementioned errors are even more successfully avoided by solution analysis in which the trace elements of interest are separated and concentrated. For many problems of extreme trace analysis, in the ng/g range and lower, this method is the only one possible since it is only after isolation and concentration of the trace elements that reliable analysis signals can be obtained.

Such procedures are easier to calibrate by standard solutions, which are, however, time-consuming and laborious and might be subject to a number of new systematic errors [1—3].

Recognition and avoidance of these errors, which appear because of losses of the element by absorption or volatilization, or because of the introduction of blank values from vessels, reagents, or dust in the air, is the chief task of extreme trace analysis today.

When it is established beyond doubt that the analytical procedure is free from such systematic errors, a principle of determination must be found which suits the procedure. This should enable the isolated trace elements to be recorded without further error, quickly and with good power of detection and reproducibility.

Everything that has so far been investigated about some excitation procedures in emission spectroscopy, the recently developed plasma sources, now give reason to hope that by these methods the above problems may be solved, although it cannot be expected that all problems will be solved optimally with these excitation sources [4—16].

(A) GENERAL REMARKS ON PLASMA EXCITATION

In addition to a few different forms of electrode such as the disc electrode, the vacuum cup electrode, or the porous cup electrode for arc and spark excitation, the flame is an important source of excitation for solutions in emission spectroscopy. Its most important fields of application are AAS and AFS since only a few elements can be determined by flame spectrometry. An important advantage, however, which may be gained by solution procedures, that of the possibility of multi-element analysis, is not possible in AAS.

As mentioned above, when using solution techniques, there is less interference than with the excitation of solid material, because matrix and inter-element effects are chiefly caused by the physical properties of solids. Only the so-called chemical matrix effects occur in solution techniques as well, such as, for example, incomplete thermal dissociation and chemical reactions in the plasma. They are dependent on the composition of the sample and the flame gases as well as on the temperature, and influence the degree of atomization. The chemical matrix effects may easily be controlled in solution procedures, or eliminated by addition of so-called releasing agents. In this way, one can avoid, for example, interference by Al in the determination of Ca by addition of Sr and La.

Combustion flames reach temperatures of about 3000 K. For OES, however, higher temperatures are desirable so that even elements with resonance lines below 350 nm are excited, and more thermally stable compounds are atomized [17,18]. As well as this, the number of emitted lines increases with temperature so that there is the possibility of avoiding spectral interference, and larger regions of concentration may be covered.

However, the higher the temperature becomes, the more marked is another chemical matrix effect, the influencing of the ionization equilibrium by the composition of the sample. Temperature, electron density, and the presence of easily ionizable elements influence the degree of ionization of an element, and through that the intensity of atom and ion lines. By addition of easily ionizable elements such as the alkali metals of the alkaline earths, this process can be deliberately influenced and controlled. This process is known as spectroscopic buffering.

The temperature and spatial temperature distribution is also an important parameter for an excitation source such as a flame, or a plasma. While a flame, the plasma of a d.c. discharge or a spark is in thermodynamic equilibrium, one must distinguish between several temperatures in other types of plasma. Atoms and ions have a certain temperature which answers to their average kinetic energy; the similarly defined temperature of the electrons is considerable higher.

TABLE 1

Some lines of interest and approximate norm temperatures [19,309]

Element	Spectral line (nm)	Norm temperature (K)
Li(I)	670.8	2850
Na(I)	589.0	6000
Mg(I)	285.0	7500
Mg(II)	280.3	16,000
Mn(I)	403.1	3700
Mn(II)	257.6	14,000
Zn(I)	213.9	11,500
Cd(I)	228.8	11,000

Such separate gas and electron temperatures are to be found chiefly in microwave-induced plasmas and hollow cathode discharges.

The intensity of an emitted spectral line, and with that the sensitivity of the determination procedure, defined as the slope of the calibration curve, is mainly determined by the temperature of the excitation source [19]. At first, this increases with temperature, falling again after a maximum value. The temperatures that give the maximum sensitivity, the so-called norm temperatures, are very high (see Table 1).

For analytical emission spectroscopy of solutions, therefore, excitation sources are required which are characterized by the highest possible temperatures. According to Boumans [17], the following come under consideration:

(1) high temperature combustion flames,
(2) current-carrying d.c. plasmas (gas- and disc-stabilized arcs),
(3) current-free d.c. plasmas,
(4) microwave plasmas with capacitive coupling (CMP),
(5) microwave-induced plasmas (MIP),
(6) high-frequency plasmas with inductive coupling (ICP), and
(7) high-frequency plasmas with capacitive coupling and special types of HF plasmas.

Since combustion flames reach only temperatures far below 5000 K, they shall not be included in the framework of this chapter. The light sources (2)—(7), above, have in part very different properties which make them more or less suitable for the purposes of solution spectroscopy.

For an introduction to the principles of high-frequency [20,21] and microwave physics [22—25], the reader is referred to the appropriate texts since such material lies outside the scope of this chapter. It is also not possible to give an introduction to plasma physics and plasma chemistry here, but again there exists a wide range of literature on the subject [7,8,26—33].

(B) HISTORICAL REVIEW [34—36]

The development of the plasma sources referred to in this chapter was begun by Gerdien and Lotz [37] about 1922. However, it was not until 1959 that the plasma jet was used for analytical purposes by Margoshes and Scribner [38] and also by Korolev and Vainshtein [39—42]. Then, from about the nineteen-sixties, there was a rapid

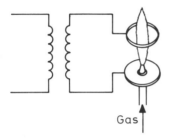

Gas

Fig. 1. Capacitively coupled HF discharge.

extension in the use of the d.c. plasma torch in many variations. However, these sources did not fulfil all requirements. In particular, the stability and susceptibility to interference left something to be desired; in some types, in fact, the emitted spectra of the electrode material proved to interfere during use. In 1941, Cristescu and Grigorovici [43–45] succeeded in reaching temperatures of 4000 K with a capacitively coupled HF plasma. This kind of plasma was then used analytically by Stolov [46] in 1956 and by Badaran et al. [47] in 1957.

The capacitively coupled HF plasma differs from arc excitation in the important particular of requiring only one electrode with a point for discharge (see Fig. 1). The plasma is induced by the strong electric field round this point. A second electrode, which is usually arranged as an annulus above the first, serves mainly to concentrate the lines of force of the electric field and only a negligible current passes through it. The two electrodes make up the capacitance in an oscillating circuit which is inductively coupled with the HF generator.

In 1960, Cristescu [48] superimposed a d.c. voltage on the HF discharge with the aid of a moveable third electrode which was mounted between the two HF electrodes. In this way, the different parameters of the plasma discharge could be controlled independently of one another.

In the years following, several types of capacitively coupled plasmas were reported [49–54]. Mavrodineanu and Hughes [55] used a HF plasma in which the power coupling on the gas flow was both capacitive and inductive, in that one end of the coil was threaded through the coil as the inner conductor and ended in a point (see Fig. 23). Over this point flamed the plasma. In 1942, Babat [56,57] published details of electrodeless discharges which

178

may be regarded as the basis of the development of ICP. Reed [58—60] in 1961 and 1962 described inductively coupled plasma torches which burned with Ar at atmospheric pressure, at first an instrument with silica tube injector with tangential gas entry, later with three concentric tubes with a centre power feed. Reed described the plasma which he used to grow single crystals, and also suggested that it could be employed as a light source in spectroscopy. He could measure temperatures up to 16,000 K. Then, in 1964, there appeared the first work on high-pressure plasmas as spectroscopic emission sources [61,62], in which Greenfield et al. compared the high-frequency plasma with inductive coupling (ICP) as a new spectroscopic light source, with the d.c.-arc-plasma jets.

Strangely enough, the first type of ICP was used with the plasma burning downwards, because at first there were difficulties, due to the construction of the apparatus, in starting the plasma with the flame upwards [63]: the connections to the HF coil in this upright position ran parallel to the plasma jet on both sides, and too much energy was diverted by the ionization of the gases flowing between the connections for a plasma to be supported. In later reports [64], these shortcomings were rectified.

The initial analytical studies with ICP were carried out independently by two teams led by Greenfield [65—70] and Fassel [71—74]. Important contributions were also made by Boumans and his colleagues. Because of its excellent characteristics, this new tool enjoyed wide popularity, as witnessed by over two hundred papers in a few years.

Parallel to the development of ICP ran that of the microwave plasma torch. Plasmas induced by microwaves may be maintained in various ways. In 1952, Broida and Moyer [75] analyzed H—D mixtures by means of an electrodeless HF discharge (150 MHz) in the gas flow, and in 1958, Broida and Chapman [76] investigated isotopes of nitrogen with a similar apparatus at 2.45 GHz. Yamamoto [77,78], and also Runnels and Gibson [79] analyzed solutions, and McCormack et al. [80], in 1965, described a detector for gas chromatography in which the organic compounds in the argon carrier gas were excited by a microwave plasma to emit various atomic lines or molecule band spectra.

Microwave plasma torches, in which the energy is coupled capacitively through a conductor with the gas (CMP), originated with Cobine and Wilbur [81]. The plasma burns above the point of an electrode in a discharge tube. In 1963, this microwave plasma was introduced into the spectral analysis of solutions by Mavrodineanu and Hughes [82]. The gas temperature reached about 3000 K while the excitation temperature lay considerably higher.

At the same time, Jecht and Kessler [83,84] reported on a similar plasma torch. The most frequently used frequency is 2.45 GHz, but others are also used, for example 520 MHz [77]. Sealed samples were also excited to emission in a low power microwave field [85—88] to reach a high power of detection in a static system. Similar electrodeless discharge lamps (EDL) were found useful in AAS.

(C) REMARKS ON HANDLING HF AND MICROWAVE ENERGY

In almost all the published work where HF or microwave energy was used, little consideration has been given to the damage to personal health, although in the various countries there are very strict safety regulations. The US National Safety Standard [89], on which the western European regulations are modelled, sets the maximum allowed microwave field strength for areas where persons work at 10 mW/cm^2, while in the USSR, the safety limit is as low as 0.01 mW/cm^2. The effect on biological tissue is still not fully understood [90—92]. It appears that the damage is not only accounted for by thermal decomposition of proteins. For example, even at 1 mW/cm^2, thermal effects on the brain and the pituitary gland, which influence the hormone balance [90], have been noted. With exposure to large doses, eye injuries (cataracts), changes in blood composition, angitis and temporary sterility have been reported. Levels of 0.1 mW/cm^2 can cause headaches and memory difficulties [93]. Especially at risk are the eyes, since, due to the lack of blood vessels, heat is not conducted from the lens, the protein is denatured and the eye becomes opaque.

Every HF and microwave generator radiates above its working frequency a wide spectrum of superimposed waves and so contributes to the electromagnetic pollution of the environment. This harmonic radiation can affect commercial communication equipment, which means that effective screening of the equipment is required to fulfil the government laws and the regulations of the boards of control (e.g. the Federal Post Office in West Germany). In the U.S.A., the generators must meet the FCC radiation specification requirements. Often, this may be achieved by relatively simple methods, especially with microwaves. Closed metal cages round the source of the radiation generally offer adequate protection if the holes and slits in the metal are small relative to the wavelength of the radiation.

Certain frequencies, e.g. 27.12 MHz, or 2.45 GHz are unlimited for industrial use. At these frequencies, the radiated power may be considerably higher than at the limited ones. If too strong harmonics are prevented by the construction of the generators, filtering and shielding is much simpler.

(D) INSTRUMENTATION

To register the analysis signals emitted by the plasma, photometers, spectrographs, or spectrometers are necessary whose optical parameters, resolving power and energy throughput add considerably to the analysis data obtained by the whole apparatus, such as precision (standard deviation) and power of detection.

The plasma excitation instruments at present on the market are, for the greater part, an integral component of the spectrometer (e.g. Spectraspan III from Techmation, ICP Plasmaspec from Kontron, the Chemical Analyser 33000 CA, or the Automatic Liquids Analyser 33000 LA from ARL). They may also be obtained, however, as a separate excitation unit. The instruments described in this text are mainly of the author's own construction, and were combined with what spectrometers were available. So the data provided in the later sections do not, with one exception [94], permit a genuine comparison of the efficiency of the various kinds of plasma.

Two optical parameters, the resolving power and the optical conductance, determine the efficiency of a spectral device. The resolving power, R, is defined as the quotient $\lambda/d\lambda$, and characterizes the wavelength difference $d\lambda$ which may be resolved at a certain wavelength λ; two lines separated by $d\lambda$ appear separately in the spectrum. The F number in prism spectrographs is determined by the focal ratio f/d of the optics where f is the focal length and d the diameter. In spectrometers with electrical reading, the radiation power at the exit slit is the decisive factor. It is determined by the optical conductance which depends on the width and the length of the entry slit and the value of $(f/d)^2$. In practice, the resolving power should be high to separate the analysis lines from interfering lines from the matrix and accompanying elements, and also from the plasma itself. In addition, the optical conductance should also allow very low intensities to be registered so that the highest possible power of detection is available [95]. This and high resolving power are, however, obtained only with very complex instrumentation, so that one must make compromises. Only Echelle spectrometers [96], which have been on the market for a while (e.g. Spectrametric, Inc., Andover, Mass.), reach this goal relatively simply. Since these spectrometers can project a two-dimensional spectrum with high resolution on a small area, they are very suitable in connection with image dissector tubes [97] as detectors. In practice, monochromators, spectrometers, and spectrographs of all types are used, and also interference filters can be used when the element of interest is separated. Photomultipliers usually serve as radiation receivers, but photodiodes and photodiode arrays are also used. The photographic plate remains an important radiation detector in multi-element analysis, in spite of its disadvantage in the cumbersome development procedures, and the properties of the photographic layer, since it can "collect" light over a long period and therefore can detect even very faint lines.

Simple monochromators are advantageous if only one element is to be determined with a low detection limit, as, for example, Hg with MIP [98]. The other extreme is represented by multi-channel spectrometers with forty or more channels, which are connected to a computer for fully automatic data recording and processing.

(For the principles of emssion spectroscopy, see refs. 99 and 100.)

(E) CLASSIFICATION AND NOMENCLATURE OF PLASMA SOURCES

Classification of different objects into groups always has its problems, since the boundaries which must of necessity be drawn are not suitable for every object. Here, also, the chosen division of the plasma excitation sources must rest on characteristics which cannot clearly be defined for every instrument.

An important feature of plasma excitation is the frequency, by which we may differentiate three groups [94].

(1) D.c. plasmas: (a) current-carrying, (b) current-free.

(2) High frequency plasmas: (a) capacitive coupling, (b) inductive coupling.

(3) Microwave plasmas: (a) capacitive coupling, (b) microwave induced.

A further possibility of classification is division according to construction. Neither method of division is in itself totally satisfactory.

The frequency for microwave generators is almost always 2.45 GHz, and the limit for high-frequency instruments must be drawn arbitrarily. By general agreement, this lies at 1 GHz, corresponding to a wavelength of 30 cm. The classification of various microwave and HF plasma torches cannot be made strictly according to this limit [77,81].

The division of d.c. plasmas, HF plasmas and ICP's appears straightforward. However, in one case the HF is superimposed on a d.c. voltage [48,101,102], in another [55,103] the power transmission is both inductive and capacitive.

The nomenclature of the various plasmas is not uniformly used in the literature. While Barnes [104] wants to use the terms "plasma torch" and "plasma flame" only for combustion processes, Greenfield et al. [36] see the descriptions suitable for electric plasma as well, since they well describe its appearance. On the other hand, they criticize the suggestion of the IUPAC that all plasma sources should be described as "plasma jets" as not differentiating enough between them. The division of plasmas in this chapter is according to that given in Sect. 1A [17]. The terms used do not strictly follow the IUPAC recommendations.

2. List of symbols and abbreviations

bar	unit of pressure (10^5 Pa)
c	centi (10^{-2})
c	speed of light (m s^{-1})
d	diameter
d.c.	direct current
eV	electron volt
f	focal length (m or mm)
g	gram
k	kilo (10^3)
l	litre
m	metre
m	milli (10^{-3})
min	minute
n	nano (10^{-9})
s	second

s	standard deviation

$$s = \sqrt{\frac{1}{n-1}\Sigma(x_i - \bar{x})^2}$$

A	ampere
AAS	atomic absorption spectroscopy
AES	atomic emission spectroscopy
AFS	atomic fluorescence spectroscopy
C	capacitance (Farad)
°C	degree centigrade
CMP	capacitively coupled plasma
DCP	direct current plasma
EDL	electrodeless discharge lamps
FAAS	furnace atomic absorption spectroscopy
FID	flame ionization detector
G	giga (10^9)
GC	gas chromatography
HF	high frequency
Hz	Hertz, unit of frequency (s^{-1})
ICP	inductively coupled plasma
IUPAC	International Union of Pure and Applied Chemistry
K	Kelvin
L	inductance (Henry)
LTE	local thermal equilibrium
M	molarity
M	mega (10^6)
MIBK	methyl isobutyl ketone
MIP	microwave induced plasma
N	normality
OES	optical emission spectroscopy
PTFE	polytetrafluoroethylene
R	resolving power ($\lambda/d\lambda$)
RF	radio frequency
UHF	ultra high frequency
V	volt ($JA^{-1} s^{-1}$)
W	watt ($kg\ m\ s^{-1}$)
λ	wavelength (nm)
λ_0	wavelength in vacuum (nm)
μ	micro (10^{-6})
ν	frequency (s^{-1})

184

3. Microwave plasmas

(A) FUNDAMENTAL PRINCIPLES [5,105,106]

Microwave plasmas, like HF plasmas, may be divided into two groups by the method of power transmission to the working gas: (a) capacitively coupled plasmas (CMP) and (b) microwave induced plasmas (MIP). The two sorts are very different in construction, efficiency, and analytical possibilities, but both have the same working frequency, which, with a few exceptions, is 2.45 GHz. This is partly due to the fact that this frequency, together with a few others, e.g. 0.461 and 5.85 GHz, has unlimited use for commerical purposes, but also because many components such as wave guides, microwave resonators and magnetrons are available for this frequency, and their cost is kept low because of the large numbers produced. With electromagnetic waves of frequencies greater than 0.1 GHz, the wavelengths are of the same order of size as the components. This means that the printed circuits generally used in electronics cannot be applied, since a high radiation loss would be the consequence.

Determined by the physical properties [20—26,28,107] of these electromagnetic waves, the dimensions of the components must be exactly tuned to the wavelength. Besides the unit of one wavelength, which at 2.45 GHz in a vacuum is $\lambda_0 = c/\nu = 2.998 \times 10^8/2.45 \times 10^9 = 12.24$ cm, the values of $3\lambda/4$, $\lambda/2$ and $\lambda/4$ are also important for the dimensions of the components [c = speed of light (m/s), ν = frequency (s^{-1})]. In <u>rectangular</u> waveguides, however, the wavelength becomes $\lambda_H = \lambda_0/\sqrt{1 - (\lambda_0/2a)^2}$ and the values of λ_H are here the critical values. (The dimension a is shown in Fig. 3).

For the generation of very high frequencies, the usual well-known components of radio and high-frequency technology, such as valves and transistors * cannot be used.

At low power, the so-called "Klystron" may be used as an oscillator; microwave generators at powers between 100 W and a few kW are driven by magnetrons [107]. These contain, in a compact form, all the necessary HF components to produce a fixed fre-

---

* The present state of transistor technology gives rise to the possibility that, by further miniaturization, transistor oscillators and amplifiers may be built using Gunn-effect and avalanche diodes, varactors etc.

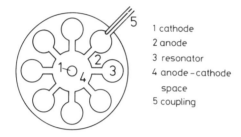

1 cathode
2 anode
3 resonator
4 anode – cathode
   space
5 coupling

Fig. 2. Cross-section of a magnetron.

quency. A schematic presentation of such a magnetron is given in Fig. 2 [23,25].

The electrons emitted by the heated cathode, 1, flow outwards to the annular anode, 2, which contains a certain number of cavity resonators, 3. Perpendicular to the plane of the paper runs a very strong stationary magnetic field. In the cavities, the UHF oscillation is excited at the lowest natural frequency so that the phases of the fields excite a travelling wave in the space 4 between the anode and cathode, which in turn links the fields in the resonators. The electrons are accelerated by the circling electromagnetic field according to the phase and return to the cathode where they knock out additional electrons and contribute to the heating of the cathode, or else they are retarded and give up part of their energy to the field.

In the process, they are focussed into "spokes" and reach the anode in irregular paths. In the spaces between the circulating "elec-

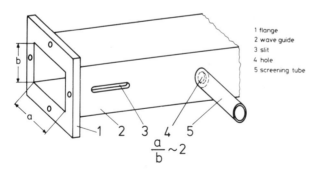

1 flange
2 wave guide
3 slit
4 hole
5 screening tube

$$\frac{a}{b} \sim 2$$

Fig. 3. Rectangular waveguide.

186

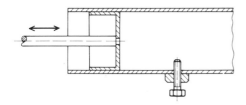

Fig. 4. Cross-section of a rectangular waveguide with tuning stub and tuning screw.

tron spokes", there are only a very few electrons. The anode current is stabilized by the current supply of the magnetron and can be regulated to control the power output [108].

The microwaves leave the apparatus at 5 and are led to the user by coaxial or cavity waveguides. Both kinds of conductor must be tuned to the wavelength, that is, be of a given length. Cavity conductors are rigid metal tubes of silver, copper, or brass with a rectangular or circular cross-section in which the microwaves propagate themselves (see Fig. 3). The dimensions of the cross-sections are standardized [23], e.g. at 2.45 GHz they are inner width = 91 mm, inner height = 42 mm. Individual guides must be linked by extremely close-fitting flanges since bad connections can result in loss of radiation. Mismatching is corrected by tuning stubs or tuning screws in the cavity waveguide (tuning transformer) (see Fig. 4).

Slits in the walls of the cavity waveguide only produce no power loss if they are arranged in the middle of the narrow side parallel to the main axis (see Fig. 3). Holes in the wall, e.g. windows for optical radiation, must be screened with the aid of a fitted tube, a so-called damping chimney in which the UHF waves are absorbed.

So-called bi-directional couplers serve to measure the power transmitted from the magnetron to the user and the reflected power. Tuning stubs and bi-directional couplers can also be used with coaxial waveguides.

If the microwave energy is to be coupled with the gas stream, a microwave, or cavity resonator is required. It is formed from a closed metal tube of rectangular or circular cross-section and the inner dimensions of the resonator must again be the same as above so that standing waves can arise.

Various types are in use [109—115], those most often employed in practice are (1) tapered rectangular type, (2) Evenson 1/4 wave

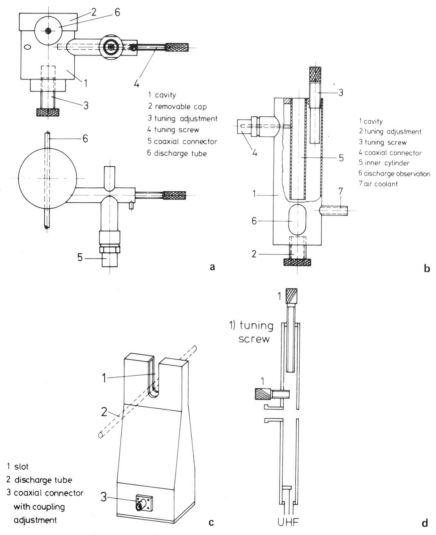

1 cavity
2 removable cap
3 tuning adjustment
4 tuning screw
5 coaxial connector
6 discharge tube

1 cavity
2 tuning adjustment
3 tuning screw
4 coaxial connector
5 inner cylinder
6 discharge observation
7 air coolant

a

b

1) tuning screw

1 slot
2 discharge tube
3 coaxial connector
   with coupling
   adjustment

c

UHF

d

Fig. 5. Microwave cavities. (a) 1/2 Wave cavity (EMS 216 L); (b) 3/4 wave cavity (EMS 210 L); (c) tapered rectangular cavity; (d) cavity after Beenakker [116].

cavity, (3) Broida 1/4 and 3/4 wave cavity and types derived from these [see Fig. 5, (a)—(d)]. The coaxial gap adjuster and the tuning stub make possible the optimal adjustment of the cavity, since the

188

plasma influences the system's impedance and also the introduction of the sample changes the plasma impedance.

The adjustment of the impedance is especially important if aqueous solutions are brought into a low-power MIP. In many cases this is not possible since H_2O and organic compounds and also foreign gases at rates of over 1 mg/s extinguish the plasma [111]. Lichte and Skogerboe [112], by changing the adjustment mechanism of an Evenson 3/4 wave cavity type, and axial arrangement of the silica tubes in the resonator, got the working gas to flow through the zone at maximum field strength. With this arrangement they could also excite atomized solutions in the plasma [see Sect. 3.B(1)].

(B) MICROWAVE INDUCED PLASMAS

(1) Instrumentation

The main components of a MIP (see Fig. 6) are the microwave generator, 1, the microwave resonator, 2, with waveguide, 3, the quartz capillary, 4, and the sample injection system, 5. Besides this, a gas unit, 7, is needed to ensure a constant gas supply, and, if the plasma is to be maintained at low pressure, a vacuum pump.

For MIP's generally, only a low power, about 100—200 W, microwave generator is needed. Table 2 lists some of these instruments with some data and their uses for analytical purposes. As an example, there is a description of a generator with the name Microtron 2002

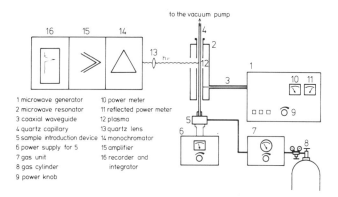

1 microwave generator
2 microwave resonator
3 coaxial waveguide
4 quartz capillary
5 sample introduction device
6 power supply for 5
7 gas unit
8 gas cylinder
9 power knob
10 power meter
11 reflected power meter
12 plasma
13 quartz lens
14 monochromator
15 amplifier
16 recorder and
     integrator

Fig. 6. Schematic diagram of a microwave induced plasma.

TABLE 2

Examples of MIP equipment

Manufacturer	Generator type	Max. power (W)	Cavity	Working gas
Monsanto Research Corp.	Mod. 207	100	Evenson $\lambda/4$	He
Scintillonics, Inc.	HV, 15 A	100	Evenson $\lambda/4$	He
Scintillonics, Inc.	HV, 15 A	100	Evenson $\lambda/4$	He
Electro-medical Supplies	Microtron 200, Mark II	200	Evenson $\lambda/4$ (214 L)	He
Electro-medical Supplies	Microton 200, Mark II	200	special	He
Electro-medical supplies	Microton 200, Mark II	200	Evenson $3\lambda/4$ (210 L)	Ar
Ito Chotampa	KTM-150	50	tapered rectangular NBS No 1	Ar
Ito Chotampa	MR III S	100	Tapered rectangular	Ar
Ito Chotampa	KTM-150	50	Tapered rectangular	Ar
Scintillonics, Inc.	HV, 15 A	100	Evenson	
Raytheon Comp.	PGM-10X1	120	Tapered rectangular, 7097-1001 G1	Ar
Opthos Instrument. Co			NBS-Rep. No 5	Ar

Mark II, Type 2001 L from Electro-Medical Supplies (Greenham) Ltd., Wantage, Berkshire, England. The compactly built instrument which measures 50 × 30 × 30 cm (approx.) works with a magnetron type JP-02 (Mullard/Philips) at 2.45 GHz. The output power is adjustable between 20 and 200 W and may be read directly from a power meter. The reflected power, which may be read from another power meter, is kept below 50 W by an automatic safety switch. At higher values, the instrument switches to "stand by", otherwise the magnetron may be damaged.

190

Pressure (mbar)	Sample introduction system	Determined elements	Matrix	Ref.
4—5.4	Ta filament	Cd, Cu, Co, Fe, Mg, Mn, Pb, Zn	Organic	117
6.3	Gas chromatograph	Cr	Trifluoro-acetylace-tonate	207
Normal	Arsine generator	As	H_2O	148
13—130	Gas chromatograph	Hg, Cl, I, ^2D, ^{12}C, ^{14}N, ^{15}N etc.	Organic	193
Normal	Gas chromatograph	C, H, S, Cl, Br, I		119
Normal	Cu net	Hg		98
Normal	Ultrasonic nebulizer, desolvation	Ag, Al, Bi, Ca, Co, Cu, Fe	H_2O	131
Normal	Arsine generator	As	Al	149
Normal	W loop	Ag, Ba, Cd, Cu, Pb, Zn	H_2O	126
	Pt filament	Ag, Co, Cr, Cu, Fe	H_2O	79
Normal	Nebulizer, desolvation	Cd, Ga, In, Hg, Zn	H_2O	132
Normal	Ta strip and carbon cup	As, Be, Bi, Cd, Co, Ni, Pb, Sb, Se, Sn, Zn	H_2O	121

The UHF energy is conducted from the exit by coaxial connectors and a flexible coaxial waveguide, 3, (50 ohm impedance) with PTFE insulation to the resonator, 2. A safety switch prevents operation of the instrument without a connected cable. The standard lengths of the available waveguides are 43, 134 and 275 cm. The instrument is controlled by illuminated knobs with a control knob for the power. The frequency of the microtron can be modulated by an additional instrument so that, for example, electrodeless discharge lamps (EDL) may be used in AAS.

A silica capillary, 4, runs traversely or axially through the cavity, the important thing here is that the working gas flows through the zone with maximum energy density [112]. The dimensions of the capillary vary. While the length is not important and is determined by the construction of the MIP, the inner cross-section is critical; the values lie between 1 mm [98] and 7 mm [117] but mostly they are 2—3 mm. Argon serves as the working gas, but in some cases He or N_2 are used. The necessary gas flow rate lies between about 50 and 1000 ml/min.

The MIP works with Ar as plasma gas at either atmospheric or reduced pressure, while with He and the conventional cavities, the plasma is only stable below 65 mbar [118] except the plasma described by Beenakker [116,119], which burns stable in a special cavity even with He at atmospheric pressure.

Since the magnetron can suffer damage if the reflected power is too high, and since practically all of the energy is reflected when the plasma is still unlit, the ignition takes place at as low a power as possible. Either a Tesla coil is used whose high voltage sparks induce in the gas stream the necessary ions and electrons to start the plasma, or a spark flash which is caused by a piece of wire which short circuits the inner tube of the resonator and the outer skin. With the aid of the tuning stub and by adjustment of the silica capillary the tuning of the system may be corrected. At optimal matching, the reflected power meter reading is at minimum. According to the power taken up, the plasma heats the silica capillary more or less strongly. A flow of air from the side cools the tube to stabilize the plasma.

The intensity of the radiation emitted by the MIP in general increases linearly with increased power input and levels off after a certain value which depends on the construction of the MIP. The intensity of a line may, however, decrease even with increasing power if, during the process, higher excited states are preferentially formed. The length of the plasma zone in the silica tube also rises with increasing power. The plasma is viewed "end on" [120] or perpendicular to the silica capillary.

The sample which is to be excited is fed into the gas stream by a sample injector system and is carried to the plasma zone by the working gas. Various sample injection systems are described in the literature but always only small quantities of sample can be introduced into the plasma per unit time; the plasma is usually

192

extinguished even by milligram quantities. H_2O and organic solvents also show this effect. Many of the analytical procedures described, therefore, are based on the evaporation and excitation of dried solutions, where microlitre quantities are injected. Carbon cup sampling devices [121,122], tantalum strip assemblies [121,123] and wire loops [124—126] of Ta, W or Pt are described in the flameless AAS type of atomizing system. Layman and Hieftje [127] developed a micro-arc sample atomizer with which the similarly dried samples are atomized. Only the MIP described by Beenakker [116,119] can be run with nebulized sample solutions without desolvation. Many elements are easily transformed into volatile compounds with chemical generation processes (e.g. BF_3, SiF_4, $GeCl_4$, $AsCl_3$, SeO_2, $SeBr_4$, SeH_2, TeH_2, AsH_3, SbH_3, BiH_3). The required reactions can take place in small vessels through which the working gas flows and transports the volatile compounds to the plasma zone. Mercury can be concentrated on an Au absorbent from air, water and other samples and, after heating, likewise enters the plasma [98], or it may be transported directly into the plasma by the method of flameless AAS after reduction to the element from aqueous solution [128].

Gaseous samples are very easy to inject directly into the working gas with the aid of gas-tight syringes through a rubber seal [129,130]. If directly coupled to a gas chromatograph, the MIP serves as detector (see Sect. 3.D).

Lichte and Skogerboe [112] succeeded in analyzing aqueous solutions with the aid of an nebulizer linked to a spray chamber heated to 160°C. Ultrasonic nebulization followed by heating and cooling of the aerosol is also described [131—133].

The properties of a microwave-induced plasma are similar to those of a CMP [111,134—139] but, due to the low power of the MIP, the CMP is superior. Electron densities are in the region of 10^{12}—10^{15} cm^{-3} and depend on the nature of the working gas, the input power, and the pressure. They are higher in the middle of the plasma than in the outer regions. The excitation processes by which the spectra in a MIP are produced are discussed by Beenakker [119].

The excitation and ionization temperatures can be ascertained from the intensity of multiplet lines from the gas. For Ar at 1 bar, they lie at 5000—6000 K. Values up to 8500 K have been reported. They drop steeply with lowered pressure; at 16 mbar they have values between 4000 and 4300 K. The gas temperatures are considerably lower, so the plasma is not in thermodynamic equilibrium.

Values between 1000 and 2500 K in the literature may be found [111,132,140,141]. Fallgatter et al. [132] and others found that gas and excitation temperatures rise if aqueous solutions are nebulized in an Ar plasma at atmospheric pressure:

without aspiration: gas temperature 1440 K; excitation temperature 8000 K.

1.1. ml H_2O/min aspirated: gas temperature 2440 K; excitation temperature 8500 K.

The temperatures are not the same in every zone in the plasma, but have a certain profile depending on the conditions of excitation. Similarly, different zones of maximum emission have been observed with radial distribution [132].

The power of detection of this plasma source varies greatly according to the construction and method of use so that general statements are not possible. With solution procedures, most values of a few ng/g can be reached. The limit to which values can be established with microtechniques lies between 10^{-10} and 10^{-13} g [see Table 3, Sect. 3.B(3)].

(2) Practice

Since the microwave-induced plasma at low power is sensitive to accompanying materials, it is usually necessary to separate and isolate the trace elements of interest from the matrix. Microtechniques are chiefly used for the introduction of the sample solution, so it is advantageous to concentrate the trace element into the smallest possible volume, to use to the full the power of detection of this excitation source. For separation and concentration, a host of possibilities from analytical chemistry are available. However, it must be ensured that no systematic errors during the chemical operations falsify the results of the analysis. The danger of this becomes greater as the trace concentrations become less [1—3].

Before beginning the analysis, the radiation path must be adjusted. This is best done with the help of one of the gas lines of the MIP. Included in this is the adjustment of the silica tubes in the resonator and the matching of the microwave system to minimum reflected power at optimum power, after the gas flow rate and then the working pressure have been set, and the plasma has been ignited by a Tesla coil. The spectrometer is then set on the desired wavelength. In general, small monochromators or spectrometers such as

194

for AAS or flame spectroscopy are used. Here there may be difficulties with individual sample introduction techniques such as the wire loop, since the sample reaches the plasma as an impulse of short duration and shows only a short signal. This time may be too short for setting up the monochromator. The situation may be helped with the light of a hollow cathode lamp or EDL which is reflected into the path of the radiation from the side.

For the actual measurement, the solution is fed into the injection system, done most easily in microtechniques with the well-known automatic micropipettes. The further course of the measurement runs in the same way as in flameless AAS. The solvent is removed by heating to around 100—120°C, then at high temperatures (depending on matrix and instrument, between 1000 and 3000 K) the solution residue is vaporized and, with the working gas, reaches the plasma zone where it is atomized and excited. The emitted light is split spectrally in the monochromator and measured in the detector. Recording the detector signal is possible in many ways, in most cases a pen recorder registers the signal as a function of time, and if, due to too wide peaks, it cannot be evaluated by the peak heights, a connected integration measures the peak areas. The evaluation of the readings can be done using a calibration curve or by the standard addition method.

Calibration curves may be set up using pure calibrating solutions. However, it must be ensured that other elements or compounds in the solution to be analyzed do not influence the analysis signal. Spectral interference and background radiation must be taken care of. Gaseous samples often offer advantages here because of lesser cross-interference and a number of elements are easier to isolate totally in this manner.

In the analysis of solutions which have passed in through a nebulizer, the plasma generally burns with a higher consumption of gas than in the microtechniques, since nebulizers obtainable from AAS and AES do not work at such low gas flows as are suitable for MIP. (The order of size is 1 l/min for nebulizer MIP's and 100 ml/min for the carbon cup [98] with an inner diameter of the silica tube of 1 mm; there are, however, reports of 2 l/min for wire loops [124]).

In general, it is expected that the limit of detection is lower if the gas flow rate is lower, because this determines the residence time of an atom in the plasma. On the other hand, after a certain value, the plasma burns more and more unstably as the gas flow becomes lower, so an optimum value must be found.

195

Another condition to be optimized for direct analysis of solutions is the temperature of the spray chamber. By heating the aerosol to about 140°C and then cooling in a water-cooled condenser, the gas is dried so that the plasma is not overloaded with the solvent.

The limits of detection which are given in the literature for such solution procedures are of the order of 10 ng/ml. The comparable values using micromethods, which have limits of 10^{-11} to 10^{-12} g [79], appear lower only at first sight. Only very small sample quantities (a few microlitres) are injected and so we lose between three and six orders of magnitude, since the preparation of solutions for analysis by macrotechniques takes place in the millilitre region. With special concentration techniques [98], it is, however, possible, for example, to separate and trace even 0.1 ng Hg from a sample of 500 mg or 10 ml of aqueous solution.

(a) Interference. The following factors influence the intensity of the emission lines of a MIP.

(1) Power absorbed from the plasma.

(2) Gas and electron temperature.

(3) Sort, pressure, and flow rate of the working gas.

(4) Sample vaporization.

(5) Sample size and composition.

(6) Observed zone of the plasma.

(1) The power taken from the plasma influences not only the intensity of the analysis line but also that of the background. This consists mainly of a weak continuum and of the lines and bands of the working gas (Ar is especially advantageous here, since its spectrum consists of only a few lines in the region of interest) and the bands of molecules and radicals. Water gives a band spectrum at above 550 nm, this region is, however, interesting in only a few cases. The intense OH band system at 306.4 nm [133] interferes more strongly, as do molecular bands of NO, NH, N_2, and carbon hydrates. The ratio of the intensities of the line and the background which represents the actual analysis signal, is maximized at various powers depending on the construction of the MIP. Together with the noise of the background, the line-to-background ratio determines the detection limit.

(2) Although the electron temperatures in the microwave plasma are high at 5500 K [141], the gas temperatures of 1000—2500 K [132,140,141] are relatively low. This means that there is too little

kinetic energy for the dissociation of thermally stable compounds. The molecules in the plasma decrease the energy of the electrons by inelastic collisions. In this way, similar conditions prevail in a low power plasma as in combustion flames, i.e. the excitation is subject to interferences as encountered in flame techniqes. According to Greenfield et al. [35], for a microwave plasma, a power of a few kilowatts is necessary to obtain sufficiently high gas temperatures. Such power is only used for a CMP jet. Lichte and Skogerboe [112], on the other hand are of the opinion that generators from upwards of 120 W power are sufficient to analyse, for example, stable monoxides, whose dissociation potential lies above 7 eV.

(3) The type of gas determines, among other things, the conditions of excitation and the spectral background. Argon, the most commonly used working gas, has good properties at atmospheric pressure. He is generally used at low pressure. Only one cavity is described for MIP operation with He and Ar at atmospheric pressure [116]. Nowadays, it seems that this He—MIP has the best properties of all. The flow rate of the working gas influences the intensity of the analysis signal. The sample is carried into the plasma by means of the gas. It depends on the speed of transportation whether the peak appears broad or narrow. With slow transportation, adsorption of the analyte at the surface of the vessel is favoured, and this leads to element losses and memory effects. On the other hand, the faster the gas flows, the shorter is the residence time of the atoms in the plasma, and the lower is the intensity of the emission lines.

(4) The vaporization of the sample is influenced by the sample introduction system in that, depending on its construction and on the constitution of the sample, the chemistry which prevails during the vaporization may be more or less complicated. For example, drying of the solution on the carrier eliminates the great advantage of solution analysis, that of having no solid sample. Also, thermally stable compounds cannot evaporate at the relatively low temperatures of the wire loop and Ta strip, and in the case of the carbon cup, some elements form very stable carbides. The best way to introduce a solution into the plasma is, therefore, nebulization, which is possible with the MIP described by Beenakker [116].

(5) Matrix effects [135] and chemical interference are observed in MIP to a lesser extent than in combustion flames [however, see (2)]. In fact, the observations of individual authors are in conflict on occasion, but this may be regarded as an indication of greatly

differing conditions of excitation of the individual plasma sources. So, far example, phosphate and silicate do not interfere with the determination of Cd [124] even at 1000 times the concentration, while the classical interference in the determination of Ca due to phosphate was also observed with MIP [112]. Elements with low ionization potential considerably influence the excitation. If the electron density in the plasma becomes higher, due to the presence of, for example, Na, the ionization equilibrium of the other elements is shifted to the side of the neutral atoms [34,127] *. A large increase in the line intensities of the atoms concerned is the result. The reports about this are numerous. The intensity of Ca emission is greatly increased in the presence of Na and K [117,131]. Manganese shows enhancement factors of 1000 in an 0.008 M solution of KCl or NaCl, and in a LiCl solution of the same strength, the Mn line is 1600 times as intense. Kawaguchi and Vallee [117] determined the excitation temperatures ** for Cu (50 μg/g) without Na at 5650 K, and in the presence of Na at 5180 K. According to Skogerboe and Coleman [111], the effects are dependent on the type of nebulizer and the aerosol facilities.

Some particulars of MIP which were not mentioned in the above outline must be noted.

The intensities of the emission lines are not the same over the entire length of the plasma; at certain points they reach a maximum. The intensity distribution can be altered by the presence of easily ionizable elements [117,142].

A strengthening of the signal with direct nebulization of organic solutions compared with aqueous solutions can be traced to a higher uptake and better nebulization because of lower viscosity and to changes in the energy characteristics of the plasma.

If the plasma burns for a longer period in a quartz capillary, a white coating on the inner wall may significantly lower the transparency of the glass [98,117]. This phenomenon is apparent especially when strongly corrosive compounds such as fluorides are present or if the tube is overheated during operation. Cooling the tube with air will not totally prevent the coating. Generally,

---

* When all reported interference of this kind is analyzed, the picture is confusing and the mechanism behind the interference appears to be somewhat more complicated.
** Rotational temperature of the R_2 branch of the (0,0) band in the system of a neutral OH molecule.

these coatings can be removed by washing for a long time in HCl or the tube must be renewed frequently. This interference will not occur if the plasma is observed from the open end of the capillary.

(3) Practical application

In order to be clear about the value of MIP for analytical purposes, it is useful to compare the advantages and disadvantages of this excitation source.

It may be considered an advantage over flame spectrometry that the excitation temperatures in the plasma are considerably higher than in combustion flames. The gas temperatures, however, are low; thermally stable compounds do not dissociate and therefore do not contribute to the emission.

A range of matrix effects and interference of accompanying components, caused by these temperature conditions, appear. The band spectra of undissociated radicals interfere. At this low power, the plasma often reacts very sensitively to larger quantities of water, organic solvents, and other materials. Easily isolated elements alter the intensity of the emission lines, often by several orders of magnitude. From these facts, if may be concluded that MIP can be used to advantage if isolated elements are to be determined which do not occur in forms that are difficult to dissociate, including gaseous samples. Also, none of the components of the sample must interfere with the determination.

Use in the direct analysis of nebulized solutions [143] appears problematic except with the MIP described by Beenakker [116]. For this, the much more powerful excitation sources of the CMP and the ICP are available. The advantages of the MIP over these instruments are to be found in the less complex instrumentation, simple construction and low costs, although when the aerosol has to be dried by the complicated method of heating and cooling, these advantages are lost again.

Some procedures described in the literature, in which, for the above reasons, MIP may be used as a meaningful instrument of determination, shall be discussed in greater detail. For many elements such as B, Si, F, P, Cr, Ge, As, Sb, Sn, Se, Te or Hg, the special advantages of a combined decomposition and separation procedure may be made use of, as they, or their more volatile compounds (halogenides, hydrides, oxides) may be evaporated into an MIP

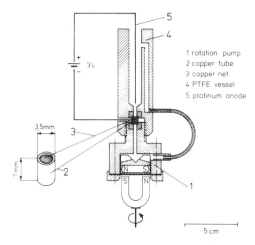

Fig. 7. Electrolysis of ng amounts of Hg^+/Hg^{2+} at a Cu cathode (by permission of the authors [98]).

directly from an electrically heated reaction vessel or after a concentration procedure.

Mercury may easily be concentrated if a gaseous sample is allowed to flow over an approximately 7 mm long column of rolled gold netting. All the Hg amalgamates on the gold [98,144]. Solutions are electrolyzed in a hydrodynamic system (see Fig. 7) in which a similar column of copper netting serves as cathode. Organic matrices may be either mineralized with HNO_3 in a PTFE bomb at high temperature and pressure, after which the Hg is separated electrolytically or else the material may be burned in microwave activated oxygen, in which the gases from the combustion pass through the Au netting column. Figure 8 shows such a system [144]. Because of the tight coupling of decomposition at 3 and excitation at 9, systematic errors may be simply eliminated in this system.

From non-volatile matrices, the Hg is driven out in a tubular oven by evaporation in the stream of carrier gas and likewise adsorbed onto gold.

When the separation of the Hg is completed, the gold or copper column may be put, after drying, into the sample introduction system of the MIP. The mercury evaporates on heating and is transported into the plasma with the working gas. The emitted Hg line at 253.7 mm may be measured with integration by means of a spectro-

200

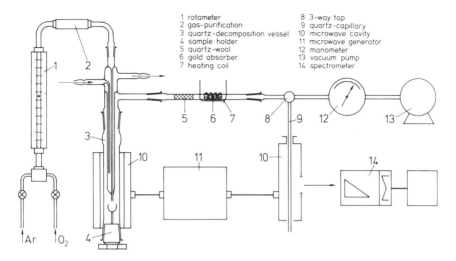

1 rotameter
2 gas-purification
3 quartz-decomposition vessel
4 sample holder
5 quartz-wool
6 gold absorber
7 heating coil

8 3-way tap
9 quartz-capillary
10 microwave cavity
11 microwave generator
12 manometer
13 vacuum pump
14 spectrometer

Fig. 8. Sampling of Hg by a gold net and excitation of the Hg in an Ar MIP after combustion of biological materials by a microwave induced oxygen plasma (by permission of the authors [144]).

photometer. However, a simple interference filter with a photo-resistive or a photovoltaic cell is also sufficient. The limit of detection of the procedure is about 0.05 ng and is sufficient to determine the ubiquitous concentration of this element even in relatively small samples (10 ml H_2O, $\leqslant 1$ g solid, 20 l air).

The sources of error in the whole determination procedure, which may easily be checked using radioactive [203]Hg, lie not in the actual determination step but mainly in adsorption losses on the surfaces of vessels, volatilization of Hg from solutions and in the blank values of vessels, reagents and Hg absorbed from the air.

Mercury may also be reduced in a vessel to elemental form as in the procedure described by Hatch and Ott [145] and driven out with Ar. The intensity of the Hg line at 253.7 nm is recorded during the period of evolution and the peak height is evaluated. The limit of detection here is 6×10^{-10} g.

A range of elements such as the noble metals, Cu, Zn, Fe, Ni, Co and others may also be separated in an electrolyzing apparatus in a hydrodynamic system in quantities of less than 0.1 ng with a yield of 95% and better in about 1—2 h. If a graphite tube is used as cathode, this can be coupled to a MIP after the electrolysis is over [146,147]

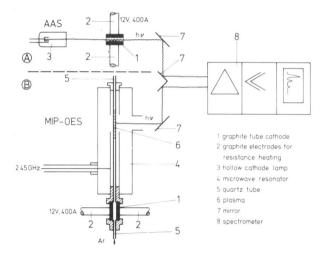

AAS

MIP-OES

1 graphite tube cathode
2 graphite electrodes for
  resistance heating
3 hollow cathode lamp
4 microwave resonator
5 quartz tube
6 plasma
7 mirror
8 spectrometer

Fig. 9. The graphite tube as a sampling system for MIP in comparison with AAS. Trace elements are separated by electrolysis.

(see Fig. 9). With electrical heating of the tube, as in the methods of flameless AAS, the separated elements are evaporated in the flow of argon and reach the plasma zone. The elements may, however, also be determined with flameless excitation in AAS. Volland et al. [147,147] compared the two methods with which they reached limits of detection of $\leqslant 0,01$ ppb.

Gases may very easily be investigated with the aid of the MIP. For this reason, this plasma source is very often used as detector for gas chromatography. A separate section is devoted to this procedure (see Sect. 3.D). Gases, however, may also be injected directly into the plasma [75,129,130,141]. Taylor et al. [130] determined the trace impurities N_2, O_2, H_2, He, Kr, NH_3, H_2O, CO_2 and organic gases in argon. The gas sample is injected with a 2 ml syringe through an injection port with a silicone rubber seal directly into the argon stream. The working gas is cleaned as the Ar passes through sulfuric acid, anhydrone, ascarite and hot zirconium sponge before the injection port. Calibration of the procedure is carried out by controlled feed rate injection of suitable gases with an infusion pump. The detection limits for the impurities lie between 0.05 and 2.1 $\mu g/g$. An exception to this is He, which can be detected only in concentrations of over 40% by weight. For the other impurities, the limits of detection suffice to analyse them in 99.9995% pure Ar without preconcentration. The authors suggest that the property of the plasma to reduce molecular impurities to potentially reactive fragments may be used to clean gases.

Serravallo and Risby [129] determined the concentration of vinyl chloride in air on the same principle with a microwave-induced helium plasma at 6.3 mbar, in which they used the Cl(II) line at 479.45 nm. As the reduced pressure discharge can only accept sample sizes as large as 100 μl, the detection limit of 390 μg/g has been obtained. As well as this, the air influences the emission of the Cl(II) line.

Lichte and Skogerboe [148] coupled an arsine generator (see also ref. 149) and a mercury reduction chamber [128] (see also ref. 150) to a MIP. For the determination of As, the 1.5 N hydrochloric or sulphuric acid sample solution passes over a column of granular zinc which has previously been washed 3 times with 0.5 ml 1 N HCl. The liberated arsine is carried out with Ar, dried with magnesium perchlorate and finally excited in the plasma. After each determination, it is ensured that the blank value remains constant by use of 0.5 ml 1 N HCl, 20—30 determinations can be carried out per hour. The limit of detection is 5 ng As.

Many authors use the MIP, when, in the manner of flameless AAS in micro-procedures, they dry small quantities of solution and inject the residue into the plasma by vaporization [117,121—126,140,151]. All the determination procedures described are similar in principle, and therefore only the Pt or W loop sample introduction device of Aldous et al. [124] will be presented as an example here.

A tube with a larger diameter is fused directly onto the quartz capillary of the plasma cell. This tube contains a platinum loop made of 10 mm lengths of 0.1 mm diameter Pt wire. If this loop forms an ellipse with major axis 0.86 mm and minor axis 0.62 mm, it picks up about 0.12 μl when immersed in a solution. A small sample vessel may be introduced into the tube from below, far enough for the loop to be dampened with the solution contained in it. By electrical heating (4.5 V, 0.1 A d.c.), the solution on the loop dries out. The plasma, meanwhile, is burning unstably. Increasing the current to 1.9 A evaporates the sample, which reaches the plasma with the Ar. The calibration curve for Cd is linear between 0.01—5.00 μg Cd/ml and the absolute limit of detection is 2×10^{-13} g. At a concentration of 0.25 μg/ml, the relative standard deviation is 10%.

TABLE 3

Detection limits of the MIP for some elements in aqueous solutions

Element	Wavelength (nm)	Detection limit (μg/ml)	Absolute detection limit (g)
B	249.8	1.0	1.2×10^{-10}
Cd	228.8	0.0017	2.0×10^{-13}
Cu	217.9	1.0	1.2×10^{-10}
Fe	248.3	2.5	3.0×10^{-10}
Hg	253.7	0.13	1.6×10^{-11}
Pb	261.4	1.0	1.2×10^{-10}
Se	204.0	3.2	4.0×10^{-10}
Zn	213.9	0.7	8.0×10^{-11}
Cu	324.8	0.01	2×10^{-11}
Cd	228.8	0.002	4×10^{-12}
Pb	405.8	0.1	2×10^{-10}
Zn	213.9	0.005	1×10^{-11}
Cd	228.8		4×10^{-12}
Co	240.7		7×10^{-11}
Pb	405.8		8×10^{-11}
Se	196.0		1×10^{-10}
Zn	213.9		5×10^{-13}
Bi	223.0		1×10^{-10}
B	249.8	0.01	
Cd	228.8	0.0004	
Co	240.7	0.06	
Hg	253.7	0.003	
Pb	405.8	0.005	
Se	196.0	0.04	
Zn	213.9	0.0006	
Cd	228.8	0.02	
Hg	253.7	0.02	
Zn	213.9	0.02	
Cd	228.8	0.04	
Cu	324.7	0.4	
Fe	372.0	1.0	
Hg	353.7	0.001	
Mg	383.8	0.1	
Se	204.8	10	
Zn	213.9	0.05	
Co	345.4	1	
Cu	213.6	0.1	
Fe	372.0	1	
Mg	279.6	0.5	
Pb	405.8	0.1	
Zn	213.9	0.01	
Hg	253.7		1×10^{-10}
Hg	253.7	0.0006	6×10^{-10}

* V = coefficient of variation.

Sample introduction system	Sample volume	Remarks*	Ref.
W loop			
Pt loop			
W loop			
Pt loop	$0.12\ \mu l$	$V = \pm 10\%$ at $0.25\ \mu l$	124
Pt loop			
Pt loop			
Pt loop			
Pt loop			
W loop	$2\ \mu l$		126
Carbon cup	$5\ \mu l$		121
Nebulizer, desolvation	1.5 ml/min	$V = \pm 1\text{--}8\%$	112
Nebulizer, desolvation	1.1 ml/min	$V = \pm 4.8\%$ at $1\ \mu g/ml$ $V = \pm 4\%$ at $1\ \mu g/ml$ $V = \pm 4.8\%$ at $1\ \mu g/ml$	132
Nebulizer, desolvation			133
Ultrasonic nebulizer, desolvation	0.5 ml/min	$V = \pm 2\text{--}5\%$	131
Hg vapor after amalgamation on Au, Ag or Cu		$V = \pm 5\%$	98
Vapor produced by a reduction chamber	$\mu l\text{--}ml$	$V = \pm 10\text{--}12\%$ for 10^{-10} g	128

In Table 3 are listed for comparison the limits of detection of a number of elements which have been obtained with the various sample introduction systems. A direct comparison of the efficiencies of the systems is, however, not possible from this table, since the parameters of the various MIPs and optical instruments used vary widely. (See also refs. 142—154).

(C) SEALED SAMPLES AND ELECTRODELESS DISCHARGE LAMPS

At low pressure, a volume of gas which is sealed in a fused glass or quartz bulb is excited to the emission of light by a HF [155] or UHF field, and a gas plasma is formed which is a spectroscopic light source. This electrodeless discharge is excited almost exclusively by microwaves, for which purpose the generators already described are used, working at the frequency 2.45 GHz [see Sect. 3.B(1)].

The electrodeless discharge lamps (EDL) are situated, held secure by means of fastening device, in the inside of a microwave cavity resonator. A flow of air takes care of the cooling necessary to stabilize the emission signal.

The EDLs have two different uses. First, they serve as a primary light source for atomic absorption (AAS) and atomic fluorescence spectrometry (AFS). Here they replace the hollow cathode lamps which, especially in the case of AFS, produce too small a light intensity which effectively influences the power of detection of this method. Secondly, the EDLs are also used as the light source in emission spectroscopy where the elements to be determined are sealed into the silica flask in a volatile form.

(1) Electrodeless discharge lamps for AAS and AFS

The preparation and separation of EDLs for AAS and AFS is described in a critical review by Haarsma et al. [115]. Although many publications concerned with the use of EDLs have appeared in the last few years, this light source has so far not replaced the hollow cathode, especially in AAS. EDLs offer great advantages for such elements as As and Se which require powerful hollow cathode lamps that are difficult to produce. However, the requirements for these elements of high intensity and stability of the emitted radiation, reproducible preparation and long working life of lamps, cannot generally all be fulfilled even by EDLs.

The production of EDLs appears simple at first glance, but many different and, in part, contradictary directions have been published and it appears that much experience and empirical knowledge is necessary for the production of good lamps. It is especially difficult to reproduce the procedure and to prepare EDLs with the desirable properties [156]. A 5—10 mm diameter quartz tube, cleaned with acid and distilled water, is sealed at one end and 15—60 mm from that end (the dimensions depend on the cavity used) is furnished with a constriction. The glass is flushed several times with a noble gas, evacuated and finally degassed by heating to as high a temperature as possible. Also the microwave discharge of a noble gas is used to clean the EDL blank. The desired substance is fed in, the lamp is again degassed, then filled with the noble gas to the optimal pressure of a few mbar, and the constriction is fused. The material introduced mostly in amounts below 1 mg, may be (1) an element (e.g. Hg, Zn, Cd, and P, As, Se, S), (2) metal halides (generally iodides or chlorides), (3) amalgams (used only in a few cases), (4) other compounds such as silicates, tetraborates and phosphates of Li, Na, K, Mg. The substance, which must be as pure as possible, must be sufficiently volatile and must not react with the quartz, neither must the dissociation products which arise during discharge.

So EDLs which are filled with gaseous hydrides of the elements Sb, Ge and Se [157] have good properties. In such a lamp bulb 1 cm in diameter and 3 cm long, with an analyte gas pressure of 1.3 mbar, about $10\,\mu g$ of the element is contained. As filler gas, He and Ar are used at an optimal pressure of 1—7 mbar.

Control of the temperature of the EDL is important in aiming at an optimal spectral output [88] since atomization processes inside the lamp are considerably influenced by the temperature. Browner and Winefordner have demonstrated [88] that in AFS with C_2H_2—air flame, the power of detection is one to three orders of magnitude lower when thermostated EDLs are used.

Multi-element EDLs can also be produced, although the amount of material is more critical. These lamps are especially interesting for AFS because they allow simultaneous multi-element analyses to be carried out, particularly since their light intensity and stability usually reach those of the single element lamps and no mutual interference among the elements results.

A wide-ranging investigation of the plasma parameters such as electric field, electron and ion current densities, electron density, ion density, electron temperature, electron conductivity and others was carried out by Avni and Winefordner [156]. Comparing the EDLs with the hollow cathode lamps, the former may be invested with the advantages of a higher light intensity with smaller background radiation, freedom from self-reversal, and a narrower line profile. It is at present, however, difficult to give universal directions about the preparation and optimum operating conditions or to make a general pronouncement on EDLs in comparison with other excitation sources for AAS and AFS, in spite of numerous publications in the past few years. For such a judgement, many mutually dependent parameters must be considered, which have been only incompletely investigated for many elements [115,158]. (See also refs. 159—170.)

(2) Electrodeless discharge lamps for OES

In most of the methods of excitation in analytical atomic spectroscopy, the short residence time of the analyte atoms in the optical path is an important limiting factor [86,155,171]. A closed system, such as that found in EDLs, in which the atoms have almost limitless residence time in the plasma, should therefore lead to a considerable increase in the power of detection. Although many attempts in this direction have been reported in the literature, so far this method has not taken hold since it has been successfully used only in a few exceptional cases, due to a lot of disturbance such as poor signal constancy, poor reproducibility of results, but especially because of the difficulty in preparing these lamps accurately and reproducibly.

EDLs pose the smallest problems when used as emission spectroscopic light sources in the investigation of gases and gas mixtures, since here the problem of evaporating the sample with the aim of keeping a constant quantity of the analyte in a gaseous state in the plasma zone is automatically solved by the nature of the procedure. So a number of investigations have been concerned with optical nitrogen-15 analysis [110,172—176], and with the determination of the $^{14}N/^{15}N$ ratio [173] in which the lamps were run at 100 MHz [173,174] and 150 MHz [110] as well as the commonly used UHF of 2.45 GHz. An advantage of this method is that very small samples, as little as $0.1\,\mu g$, are sufficient for analysis. The technique of preparation of these EDLs is similar to that used in producing lamps for AAS/AFS.

The spectral properties of a microwave helium discharge were investigated by Garber and Taylor [177]. Agterdenbos and co-workers

208

[85—87,171,178] reported on the trace analysis of Cd, In, Tl, Pb, Hg, Se, Te and Zn by microwave excitation of sealed samples. They could determine these elements with a good power of detection in the region of 4 pg to 25 ng. The reproducibility at the 5—10 ng level was about 30%. However, they did not investigate the interferences due to the matrix and accompanying elements, as encountered in the analysis of actual samples. Since a number of difficulties may be expected from this direction, this method of use for EDLs appears limited.

In the preparation of the lamps [86], the silica vessel, with a diameter of 10 mm and length of 25 mm was first degassed by heating in a vacuum. Then, 100 μl of the analyte solution and desired additional material were fed in and the solution dried out by cooling with liquid nitrogen under vacuum. Afterwards the bulb was filled with He at the desired pressure (about 7 mbar) and sealed.

There was interference due to intense band spectra [86] which stemmed from Cl, air, OH, C_2 and metal monochlorides. Addition of "getters" such as Ge or Bi suppressed these bands. During combustion, however, quantities of these metals could sublime onto the walls of the quartz bulb, where the coating that built up absorbed part of the light.

Oxidation processes in the lamp, which, for example, with Tl cause a decrease of the signal—background ratio, can be suppressed by using hydrogen as a filler gas. Gas pressure and power supplied influence the emission signal, so that optimal values for this must be ascertained. The kind of quartz glass used can also influence the discharge, the choice of a suitable kind is, therefore, of decisive importance [87]. Eckert [155] reports on initial experiments with HF-heated electrodeless discharge lamps. The power of a 25—27 MHz generator at levels of from 250 to 700 W was coupled to the lamp by a three-turn inductor coil. The discharge bulbs of quartz (diam. 35 or 46 mm) were cleaned before feeding in the sample by generating an intense induction discharge at an Ar pressure of 2 mbar. Afterwards, they were rinsed through with Ar, and the sample injected against the Ar flow. After addition of methyl alcohol, the sample was dried by heating in a vacuum. After four hours at about 10^{-9} bar, the bulb was filled with the plasma medium (Hg, Ar or Xe) and sealed. The best results were achieved with 46 mm lamps with 270 mbar Xe.

The power of detection of this method is better than comparable values for ICP—OES. For the elements Pb, Hg, Sb and Cd, it lies in the lower ng region or lower. However, even with the HF-excited sealed samples, considerable difficulties have yet to be overcome, lying mainly in the stability of the discharge, and the sample adsorption by the walls of the tube. (See also refs. 179—184.)

(D) MICROWAVE EXCITED EMISSION SPECTROMETRIC DETECTORS FOR GAS CHROMATOGRAPHY

While, until now, there have been only a few attempts to use a HF plasma (ICP) as emissive detector in gas chromatography [185—188], the microwave-induced plasma has grown more and more important [15,189—198] since 1965 when McCormack et al. [80] used it for the first time for this purpose. The construction of such a GC—MIP instrument combination is simple: the outlet of the separation column of the gas chromatograph is connected directly with the plasma capillary of the MIP (see Fig. 10) so that the effluent from the GC column passes immediately into the noble gas plasma, there to be dissociated and excited. The carrier gas of the gas chromatograph is also the working gas of the MIP. The atoms and molecule fragments emit their characteristic radiation which, as described in Sect. 3.B(2) on MIP, is detected in a small monochromator or spectrometer.

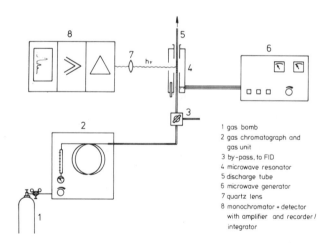

1 gas bomb
2 gas chromatograph and
  gas unit
3 by-pass, to FID
4 microwave resonator
5 discharge tube
6 microwave generator
7 quartz lens
8 monochromator + detector
  with amplifier and recorder /
  integrator

Fig. 10. Schematic diagram of a gas chromatograph with MIP detector.

210

Both argon and helium may serve as carrier and plasma gas. Both possess different advantages and disadvantages. The low wattage Ar plasma works both at atmospheric and reduced pressure. The atmospheric-type detector is easier to construct and less limited in application and selectivity [199]. However, it shows up a large proportion of organic molecules and radicals in the plasma which give a heavy spectral background by their band systems [200]. It is possible that, in such a plasma source, the atomic lines of a few elements do not show up in the spectrum as has been observed with S, Cl and Br [201]. In a low-pressure plasma, on the other hand, they are excited. Bache and Lisk [202,203] report that the operation of the plasma at reduced pressure resulted in a tenfold increase in selectivity and sensitivity of the detector for residues of organophosphorus insecticides. Also, it tolerates larger amounts of organic matter [204].

With helium as working gas, a plasma can be maintained at reduced pressure under 65 mbar [118], or with the cavity described by Beenakker [116,119] also at atmospheric pressure. Helium has the advantage compared with Ar of a lesser spectral background, since its emission spectrum consists of few lines [80,191], as well as having a higher excitation energy. Beenakker [119,205], with help of his cavity of novel design [116], developed a MIP detector with considerably improved properties. This cavity makes it possible to maintain a He plasma even at atmospheric pressure, and so circumvent the disadvantage of an atmospheric Ar plasma, which is that various elements such as F, Cl, Br, N and O emit only their diatomic band spectra, while in the He plasma, the elemental emission lines appear.

Analysis signal and background noise are only slightly dependent on the microwave power and the flow rate. A comparison of the limits of detection obtained with this detector with the atmospheric Ar-MIP and the reduced-pressure He-MIP shows an increase in the power of detection by one or two orders of magnitude. Moye [206] used Ar—He mixtures in the determination of organic phosphate pesticides and observed an enhancement of sensitivity and selectivity.

If the gas chromatograph and MIP are closely coupled, a problem arises in the optimization of the gas flow rate for the plasma, since the intensity of the emission lines is greatly influenced by this. However, the GC column also has an optimum value of gas flow rate for the best resolution. The problem may be overcome by splitting the

211

TABLE 4

Some detection limits for MIP–GC

(A direct comparison of the data given is not possible as they are obtained under different conditions.)

Compound	Detected element	Analysis line (nm)	Detection limit ($g\ s^{-1}$)	Plasma gas	Pressure (mbar)	Ref.
Organic	C	247.9	8×10^{-11}	He	>0.3	195
	H	486.1	3×10^{-11}			
	F	685.6	6×10^{-11}			
	Cl	479.5	6×10^{-11}			
	Br	470.5	9×10^{-11}			
	I	516.1	5×10^{-11}			
	S	545.4	9×10^{-11}			
Organic	C	193.1	4×10^{-13}	He	Atmospheric	119
	H	486.1	2×10^{-12}			
	Cl	479.5	7×10^{-12}			
	Br	470.5	5×10^{-12}			
	I	516.1	3×10^{-12}			
	S	545.4	2.5×10^{-11}			
C_6F_6	C	516.5	3×10^{-14}	Ar	Atmospheric	80
	F	516.6	3×10^{-12}			
$CHCl_3$	Cl	278.8	8×10^{-10}			
CH_3Br	Br	298.5	2×10^{-7}			
CH_3I	I	206.2	7×10^{-14}			
CS_2	S	257.5	1×10^{-9}			
Organophosphorus insecticides	P	253.6	$\sim 5 \times 10^{-12}$	Ar	Atmospheric	203

Substance	Element	Wavelength (nm)	Detection limit	Gas	Pressure	Ref.
Organophosphorus insecticides	P	253.6	6×10^{-13}	Ar	267	202
	C	247.9	1.9×10^{-10}			
	Cl	256 [a]	4.5×10^{-9}			
	Br	292 [a]	2.5×10^{-9}			
	I	206.2	1×10^{-10}	Ar	Atmospheric	191
	S	182.0	4×10^{-11}			
	P	253.5	3×10^{-11}			
O$_2$		Measurement of	3.8×10^{-9}			
N$_2$		the reflected	2.8×10^{-9}			
H$_2$		microwave power	0.3×10^{-9}			
CO$_2$			4.2×10^{-9}	Ar	Atmospheric	191
CH$_4$			1.0×10^{-9}			
Acetone			2.1×10^{-9}			
Ethanol			1.3×10^{-9}			
Toluene			1.4×10^{-9}			
Trifluoro-acetylaceto-nates	Al	396.2	1.9×10^{-11}			
	Cr	357.9	3.6×10^{-12}			
	Cu	324.7	8.0×10^{-12}	Ar	Atmospheric	212
	Fe	344.1	1.3×10^{-11}			
CH$_3$HgCl	Hg	253.7	5×10^{-14} g [b]	He	~80	193
CH$_3$HgCl	Hg	253.7	4×10^{-12} g [b]	Ar	6–13	204
Piazselenol	Se	204.0	4×10^{-11} g [b]	Ar	Atmospheric	226
Alkylarsines, AsH$_3$	As	228.8	2×10^{-11} g [b]	Ar	Atmospheric	223

[a] Band.
[b] Detection limits of the element.

References pp. 278—293

213

gas flow and fitting a capillary restrictor in the gas stream for plasmas, running at reduced pressure, to throttle the gas streaming to the plasma. The unused part of the carrier gas can be fed to a second detector, e.g. a flame ionization detector (see Fig. 10) which complements a MIP detector working on specific elements by showing up all the organic compounds.

As stated in Sect. 3.B, a MIP can be overloaded even with less than 1 mg organic material, so that the plasma is subject to strong interference or is extinguished. This problem appears especially when small quantities of organic materials are to be analyzed in an organic solvent and the solvent peak appears before the compound of interest. There are several possibilities of remedying the situation [191], e.g. the plasma is not lit until the solvent has been eluted [80], the solvent by-passes the detector or is back-flushed if it is eluted after the substance of interest.

Large quantities of organic materials, e.g. solvents, give rise to a further disturbance. Carbon deposits form on the wall of the plasma capillary absorbing part of the light and increasing the background of the light source [200]. This may also be prevented if the plasma is lit only after the appearance of the solvent peak [203], or if traces of N_2 [196], O_2 [195,200,207] or air [80] are added to the plasma gas. Oxygen also reduces molecular emission [200]. However, due to this addition of gases, the spectral background is considerably increased.

A brief description of the model MPD 850 (Applied Research Laboratories) is given here as an example of a commercial GC instrument with microwave plasma detector. The effluent of the gas chromatograph is fed in two similar gas streams to a flame ionization detector and to the MIP. The microwave generator with the usual working frequency of 2.45 GHz uses, at maximum, 200 W and couples the UHF power over a coaxial waveguide and $\lambda/4$ cavity resonator to the He gas flow in the 1 mm capillary. The grating spectrometer with a wavelength range of 190—800 nm (1st- and 2nd-order; radius of curvature of the grating 0.74 m; 960 l/mm) can be equipped with a maximum of 15 fixed exit slits and photomultipliers and a maximum of 4 amplifier channels so that the determination of the empirical formula of unknown compounds is possible. The limit of detection for most elements is 0.1—1.0 ng/s.

The MIP can act as a GC detector in two different ways. With selective detection, one element line of heteroatoms or substituents of organic molecules such as N, P, S, Cl, Br, F or of metals such as

Be, Al, Cr, Hg, Sc, In, Ga, Fe, Cu, Mn and V and others [208,209] in organometallic compounds or metal chelates is measured. With non-selective detection, C lines (e.g. 247.9 nm [199]) or bands (e.g. C_2 385.2 or 516.5 nm) are shown. While the detector in the first method of measurement sees only such compounds that contain the required element, in the non-selective method of working it registers all organic compounds.

Gas chromatography with MIP detector is especially suitable to register metal chelates specifically and with a very good power of detection [200,207,210—212]. The plasma detector can here replace the very sensitive electron capture detector, with which it is not possible, except by the retention time, to distinguish between metal chelates and reagents which often leave the GC column badly separated. If the reagent is present in excess, as is normally the case, the chromatogram often only shows the analyte metal chelate as a shoulder on the main peak. The MIP detector, on the other hand, will register only the metal chelate with the selective detection method.

The coupling of GC with MIP is also very advantageous for mercury determination, since in this way it is possible to record Hg in its various organic compounds, for example in food or environmental control [193,204,213—216].

In the first instance of use the MIP—GC detector, McCormack et al. [80] investigated organic compounds both with the selective and the non-selective procedures. In this process, because of poor spectral resolution, they had a lot of interference due to line or band overlapping of CN, C_2 and CH emission and others, as the emission spectra of all organic compounds contained many complex bands. The authors turned their attention to the use of the detector for F, Cl, I, S and P compounds.

Bache and Lisk [114,201—203,216—220] developed the GC—MIP technique in the following years. At first they determined organophosphorus insecticide residues in the ng/g range, then ng quantities of organic halogen, sulphur and phosphorus compounds and organic mercury compounds in a helium plasma. Dagnall and co-workers [118,190,191,199,212,221,222] have also reported, in many publications, the use of this detector. They determined organic sulphur compounds with a He plasma at 13—40 mbar. However, in the process they discovered that the identity of the compound influences the sensitivity of an emission line [218]. The atmospheric pressure Ar plasma detector was used in the conventional gas chromatographic analysis of a range of O, N and halogen-containing compounds [191] and they determined quantitative inter-element ratios of Cl, Br, I, S and P to C in various compounds, e.g. of Cl to C in carbon tetrachloride, chloroform, tetrachloroethylene, dichloromethane, trichloroethylene, dichloroethane and ethyl chloride [222]. In addition to the determination of CO, CO_2, NO, SO_2 in air [190], they also used the microwave-excited emissive detector for the determination of various metal chelates of acetylacetone and trifluoroacetylacetone [212]. This determination of metals forming volatile organic compounds which can pass through the GC column without dissociation, is the chief area of use of the GC—MIP detector.

Talmi and Bostik [223] thus determined alkylarsenic acids in pesticides and environmental samples after reduction with $NaBH_4$ to the corresponding arsines which were collected in toluene at $-5°C$ and separated on a GC column. The emission intensity of the As 228.8 nm spectra line was measured. As and Sb were also determined gas chromatographically in environmental samples after co-crystallization with thionalid and reaction of the precipitate with phenylmagnesium bromide as triphenylarsine and stibine [224] (see also refs. 225). With 5-nitro-o-phenylenediamine, selenium forms a volatile piazselenol compound [226] which is fed into the GC column in toluene solution. The maximization of the selectivity and sensitivity of the microwave-induced emissive spectrometric detector for the determination of chromium-β-diketonates by oxygen as a doping gas was studied by Serravallo and Risby [200,207] (see also ref. 210). Kawaguchi et al. [211] successfully tuned the detector to acetylacetones of Al, Be and Cr and Sakamoto et al. [227] did so to beryllium trifluoroacetylacetonate. Methylmercury chloride may be determined in water samples [204], fish [216] and environmental samples [215] with the GC—MIP detector with the same degree of sensitivity as dimethylmercury [213] and other organic mercury compounds.

The performance data obtained for the analysis procedures by the various authors are listed in Table 4. (See also refs. 228—232.)

(E) CAPACITIVELY COUPLED MICROWAVE PLASMAS

(1) Instrumentation

Side by side with the MIP, the capacitively coupled microwave plasma (CMP) represents the second type of microwave-excited

216

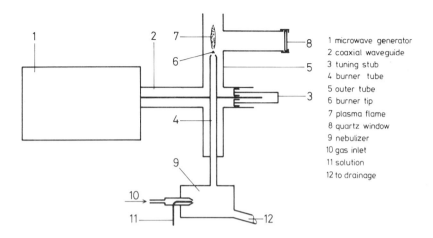

1 microwave generator
2 coaxial waveguide
3 tuning stub
4 burner tube
5 outer tube
6 burner tip
7 plasma flame
8 quartz window
9 nebulizer
10 gas inlet
11 solution
12 to drainage

Fig. 11. Schematic diagram of a capacitively coupled microwave plasma.

torch. Figure 11 shows the principal features [83,94] of such an excitation torch. It was developed by Cobine and Wilbur [81] and by Kessler and co-workers [83,84,233—235] and similar types were investigated by Mavrodineanu and Hughes [55,82], Murayama and Yamamoto [77,142,237—239], and Goto et al. [240].

The powers of the generators described are considerably higher than those of the MIPs and lie between 0.5 and 3 kW. The frequency is the same at 2.45 GHz.

The magnetron and power supply are placed in the microwave generator with the necessary controlling devices. A waveguide system 2, couples the UHF energy with the inner burner tube, 4. Either a cavity or a coaxial waveguide serves this purpose. The latter, however, is generally built not as a flexible cable but as a rigid system because the power transmission is so high. One or more tuning parts ensure optimal impedance matching.

Figure 12 shows a cross-section through a CMP burner, as distributed by the firm Erbe-Elektromedizin in Tübingen, Federal Republic of Germany. Similar systems are produced commercially by the firms A.R.L., Ecublons-Lausann and Hitachi, Ltd., Tokyo, Japan (Mod. 300 UHF Plasma Spectra Scan). The UHF energy is conducted by the cavity waveguide, 3. The coaxial electrode forms the

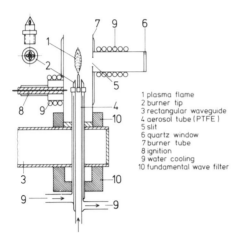

1 plasma flame
2 burner tip
3 rectangular waveguide
4 aerosol tube (PTFE)
5 slit
6 quartz window
7 burner tube
8 ignition
9 water cooling
10 fundamental wave filter

Fig. 12. Cross-section of the CMP.

actual burner. Its tip, made of gold [94] *, forms a capacitance against the earth and transfers the energy onto the gas streaming through the tip. The burning plasma can thus be interpreted as an extension of the inner conductor [83,84].

This electrode presents a source of error which is, however, mentioned as a disadvantage by only a few authors [241]: a quantity, be it ever so small, of the electrode material contaminates the sample and the lines appear in the spectrum.

The position of the tip in the outer tube, 6, is critical, since it must be sited at the point where the electric field strength is maximum. The working gas, in most cases argon or nitrogen, although air, He or Ne are also used [83], contains the nebulized sample. It flows through a PTFE tube, 4, inside the central conductor and reaches the plasma zone through drillings in the burner tip.

In some set-ups [239,242,243], the working gas and aerosol are led outside the central conductor, where a PTFE cylinder prevents the gas from getting into the rectangular cavity waveguide. This type of aerosol conduction is, however, not as effective as the one previously described, since the aerosol in part flows past the plasma

---

* Other materials such as Mo [83,233], W [233], or Al [77,237,239] are also used.

218

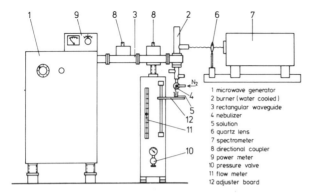

Fig. 13. Schematic diagram of a CMP apparatus (Erbe, Electromedizin, Tübingen, Federal Republic of Germany).

and does not completely reach the inner zones. A further concentric tube, 9, leads cooling water to the burner tip, which would otherwise corrode badly, and also contributes to the stability of the plasma.

The plasma is ignited by a spark discharge which produces the necessary first electrons and ions. During operation, the outer burner tube, 6, also heats up strongly and must be cooled. The light leaves through a vertical slit in this tube; a moveable second tube is arranged over the outer conductor and forms a T with a further tube at the end of which is the silica observation window, 5. By this arrangement one can observe every desired zone of the plasma, and yet the outgoing microwave radiation is so much damped that it does not exceed 10 mW/cm^2. The groundwave filters, 7, are also there for the latter reason. They are at the junction points between the burner tube and the cavity waveguide.

In Fig. 13, the whole construction of the spectrometric device is schematically presented. Between the generator, 1, and the burner, 2, is the directional coupler system, 8, with which the energy taken up by the plasma can be displayed according to choice either as the power flowing in and out or as the difference of these two quantities. To protect the magnetron against too much reflected energy, there is a damping device in a certain place inside the cavity waveguide. It is made of a glass U-tube whose length is $\lambda_H/4$ and through which water flows.

Capacitively coupled plasma torches are, without exception used for direct solution analysis, for which they are well suited thanks to the higher power, of up to a few kW, and the fact that larger quantities of solvent do not affect the plasma very much. The nebulizers used in flame techniques also serve as nebulizers here and are operated in a CMP with all the working gas. Not all types are suitable for this plasma. For example, some of the constructions used in AAS require quantities of gas which are too large for the microwave plasma [244—246]. However, the gas flow rates are higher than in the MIP, lying between 4 and 7 l/min. Kniseley et al. [246] describe a special nebulizer for small gas flows which is suitable for the CMP as well. Another one with a very low gas consumption is that of Meinhard [245].

The situation may also be improved by splitting the gas stream [94,247] and using only part of it for the plasma. However, larger quantities of the sample solution are then required and a special microtechnique, the so-called impulse technique, cannot be meaningfully used. In this, small quantities of solution between 10 and 500 μl are added via the intake capillary and nebulized [248]. Coupled with the necessary concentration procedures, this technique leads to better powers of detection.

In pneumatic nebulizers, the sample uptake rate depends to a great extent on the gas flow quantities and the length of time the atoms spend in the plasma is also influenced by it. The intensity of the spectral lines therefore passes through a maximum with increasing flow rates. It is this maximum that must be recorded. What has already been said for the MIP also applies here.

A number of authors [53,54,131,249,250] use ultrasonic nebulizers, which, when compared with the pneumatic ones, lead to different results. The decision on which type of aerosol production brings better powers of detection can obviously not be generalized. It is a disadvantage that larger quantities of solution are required and it takes a good deal longer after spraying a solution to wash the nebulizer of its residue.

The advantages of ultrasonic nebulization lie in the formation of finer and more homogeneous aerosol particles and more material is conveyed to the plasma. In this way, the gas stream does not contribute to spraying the solution, but only transports the aerosol into the plasma. Hence, for optimal adjustment of the flow rates, no compromises are necessary here. Plasma and nebulization do not influence each other.

The spectroscopic properties of the CMP surpass those of the MIP; the most important difference is that solutions may be introduced in aerosol form without particularly affecting the stability.

Because of the higher power consumption, the gas and excitation temperatures approach each other [84,251]. The excitation temperatures, however, still lie above the gas temperatures and the plasma is not in thermodynamic equilibrium [84,111,237,241].

According to Jecht and Kessler [84], the CMP does not operate on purely thermal excitation. The mechanism preferentially proceeds by electron excitation or by step processes with metastable states [5].

The temperature data in the literature are dependent on the plasma parameters. The maximum gas temperatures lie between 3000 K [55], 4500 K [111] and 5700 K [242,243] while the excitation temperatures vary between 4900 K [242,243] and 8150 K [77]. The temperatures rise with increasing ionization potential of the working gas [111]. When compared with MIP, the excitation temperatures are nearly the same while the gas temperatures lie considerably higher, as they increase with power. For the noble gases such as He and Ar, the gas temperatures remain low at around 1000 K while molecular gases such as N_2 reach several thousand K. Here, the electrons transfer part of their energy to the molecule by exciting vibration and dissociation [5]. Thus by collision and recombination, the gas temperature is raised at the expense of the electron temperature.

In the description of the temperature distribution, the various plasma zones which form both radially and vertically must be taken into account [83]. Murayama [237] differentiates up to five such zones, in which different excitation conditions exist for the various elements. Inside an argon plasma, in the "core", the state approaches, with increasing power, the local thermodynamic equilibrium. If, however, sodium ions are introduced into the plasma, the excitation and ionization temperatures fall and the core departs from the local thermodynamic equilibrium. In the outer layers of the plasma, the temperatures are lower. According to Murayama the gas temperatures here are not dependent on the power.

Kitagawa and Takeuchi [242,243] have also established temperature profiles in Ar and N_2 by wide-ranging measurements. While the electron temperature of Mn atoms in a N_2 plasma falls relatively steadily from 3750 K in the core to 2700 K in the outermost zone, in an Ar plasma, the fall in temperature is steep only in the outer-most layer. A wide zone is at constant temperature, then from the level of the burner tip at 6100 K, the temperature falls to 3000 K at a height of 30 mm above the tip. These values are for a power of 250 W. In general, the temperature is highest around the burner tip, and decreases upwards and outwards. Higher temperatures also mean increased ionization with weaker atomic lines and stronger ion lines. While elements with a high ionization potential show little or no mutual interference with other elements, those with low ionization potential greatly influence the emission [94,242,243]. In CMPs they alter the excitation temperature and the state of the ionization equilibrium. In the presence of, say, Na, three situations are possible.

(a) The intensity of emission lines of elements or ions rises with increased Na concentration.

(b) The intensity falls with increased Na concentration.

(c) With increasing Na concentration, the intensity passes through a maximum.

If temperature and degree of ionization are low, changes in the chemical composition of the plasma have more effect than at higher temperatures. In this way, the interferences of the excitation in the various plasma zones have different effects. Thus, for example, As, B and Hg react less to an addition of sodium, since they are excited chiefly in the inner zone, than elements such as Cd, Al, Ċa, Ba, Sr which show maximum intensity in the outer zones [239].

The electron density of the capacitively coupled plasma is comparable with that in a MIP, having a value between 10^{12} and 10^{15} cm^{-3} [84,142,151,239,251,252]. It is dependent on operating pressure, support gas identity and input power.

The spectral background in a CMP is low with Ar. Here, only lines and bands from the solvent (with aqueous solutions, usually only OH bands) appear beside the Ar lines and the continuum. With nitrogen, the various band systems may interfere. The background here decreases about two orders of magnitude in the wavelength region from 340 to 600 nm. In the choice of the best line for analysis, no general information can be given, since for this the parameters of the plasma and the particulars of the analysis problem must be taken into account. Table 5 (p. 227) shows the limits of detection at the wavelengths of the lines used and may serve as a first orientation.

The intensity data given in the spectroscopic tables are not always valid for plasma excitation, depending on the excitation conditions.

Other lines or even ion lines may be more intense. Since the background is not constant over the whole spectrum, a less intense line may have a better limit of detection because of a better signal-to-background ratio or a lower background noise. Possible interfering bands must also be considered in the selection of the analysis line.

(2) Practice

Although there were various attempts to introduce solid samples into a CMP [234], its chief area of use is almost totally confined to the excitation of nebulized solutions. Due to interference by easily ionizable elements, it is used to greatest advantage in cases where the analysis problem in itself requires separation and concentration of the trace elements of interest, provided the interference from the matrix is too great in other determination procedures also or the concentrations are too low. The solutions obtained after the analysis process can then be excited with the CMP for multi-element analysis.

The preparation for the measurements, apart from the preparation of the sample solutions, is the same as those described for MIPs [see Sect. 3.B(2)] and will merely be summarized here.

(1) Ignition of plasma and optimization of the power by impedance matching.

(2) Adjustment of the radiation path.

(3) Setting of the working pressure and gas flow rate for the nebulizer, and therefore also for the plasma.

(4) Measurement of background with blanks.

(5) Measurement of the sample.

The evaluation of the measured signals is carried out, depending on the equipment, using the peak height with a pen recorder, or the peak area if an integrator is connected. Spectrometers with several fixed channels print out the measured signal, electronically integrated with respect to time. The calibration of the determination, as in all solution procedures, is generally without problems, since it can be carried out with standard solutions. All the usual techniques, such as calibration curves, standard addition or comparison with a standard may be used. If interfering elements such as the alkalis or alkaline earths are present, their concentration must be taken into account in the calibration solutions as well. If the amounts of these interfering components of the solution are unknown and effect an increase in the intensity of the lines, it is in many cases possible to buffer the interference by adding an excess of the elements in question into the calibration and analysis solutions. The limit of detection of the CMP is naturally dependent not only on the plasma source but on the parameters of the spectrometer as well, e.g. optical conductance and resolving power, the properties of the photomultiplier, the selected line, etc.

The limits of detection given in the literature lie in the region of a few ng/g, sometimes even below that [50,111,239], while some authors reach only the μg/g region [50,240,247,253,254].

(a) Interference. The most important source of interference, which in a CMP may alter the line intensities by several orders of magnitude, is the presence of elements with low ionization potential [50,94,111,142,237—239,242,243,247]. The alkaline earths have a smaller effect than the alkalis, and here Na is the most important since it occurs so frequently. These elements, by raising the electron density, affect the ionization equilibrium, which moves to the side of the atoms, and the excitation temperature falls. The processes are the same as in MIP [see Sect. 3.B(2)(a), point (5)]. The extent of this effect depends on the ionization energy of the elements to be analyzed as well as on the type and concentration of the easily ionizable element.

Kitagawa and Takeuchi [242,243] investigated the influence of the alkaline and alkaline earth elements on the excitation of Mn and Murayama et al. [142,237—239] that of Na on a range of elements. In particular with the rare earth metals, enhancement factors of 10—1000 were found. Caesium also improves the limit of detection of many elements such as Ba, Li, Al, Ga, V, Ti and others, by up to one order of magnitude [94,247]. In addition to raising the sensitivity by this matrix effect, Cs also lowers the intensity of the background in a CMP by about one order of magnitude. Thus for atomic lines, the same power of detection may be reached with the CMP as with the ICP for many elements, but the ICP is still better when using ion lines.

Alkaline metals also alter the geometric dimensions of the flame and influence the spatial distribution of the analysis elements in the plasma [94,111,237,247], which causes problems in the projection of the zones on the spectrometer slit. Since the elements have the maximum intensity of their spectra lines in given regions of the plasma, heavy interference is caused by this and the interference increases from the core to the outer layer.

The introduction of the sample into the plasma, a further

geometric factor, also plays an important part. The best method would be sample injection through the central core of the plasma [111] as happens in the ICP.

Interelement effects due to other elements are small in comparison with the alkalis [50,223,236,247,253,254]. Mutual interference obviously depends on the height, or rather the difference in ionization energies of the elements concerned. The interference mechanism is, however, complicated and at present is not fully understood, as shown by the confused picture that emerges from the literature. Only a few examples will therefore be mentioned.

While manganese is scarcely affected by cadmium and calcium, aluminium interferes strongly. Large quantities of Fe (<10 mg/ml) increase the intensity of the atomic line of Mn at 279.8 nm and weaken the normally more intense ion line at 257.6 nm [253], i.e. they alter the ionization equilibrium. Co, Cu, Ni, Pb, and Sn have no effect in concentrations between 5 and 50 μg/g, and the effect of Cr, Al, As, Mo, Sb, Tl, V and Zn is only slight. In the presence of 0.5 mg Fe/ml in 3 M HCl, no interference from any of the above elements up to concentrations of 1 mg/ml is noted. The line intensity of Cr at 259.35 nm is increased by Al, Co, Cu, Ni, Mn, Pb and Zn and decreased by Ti and Mo. Here, again, 2 mg Fe/ml in 3 M HCl eliminates this interference [254]. Copper affects the intensity of the Ni line [236].

The reports about interferences due to anions are not in agreement. While many authors observe little or no difference in intensity, Atsuya [254] (with chromium) and Atsuya and Akatsuka [253] (with Mn) established a falling off in intensity of emission of up to 50% with increasing quantities of HCl and HNO_3. Then, from 2 to 3.5 M acid, the line intensity remained constant. The depression due to H_2SO_4 was considerably greater. Goto et al. [240] found no dependence of the line intensity on acid concentration with Sn and Bi. With Al, they observed a loss of intensity of about a third. The background showed no change for HCl, HNO_3 and H_2SO_4 in the concentration range between 0 and 6 N. According to Gebhardt and Horn [235], the anion concentration has a detectable effect on the intensity of the spectral lines. With larger amounts of anion concentration, differences are less noticeable.

Increase in intensity due to organic solvents [236] can certainly be traced, to a large extent, to the more effective nebulization and higher rates of uptake than with aqueous solutions, as is also observed in flame techniques.

(3) Practical application

Although the main field of use of the CMP is in trace analysis, due to its properties and good limits of detection, in the early beginnings main and subsidiary components were also determined using this excitation source. Thus Kessler and Gebhardt [233] used the microwave plasma to good effect for speedy and accurate determinations of the subsidiary components Mg, Al, and Fe in limestone, and Al and Fe in dolomite. After decomposition of the samples, the solutions could be sprayed into the plasma. The calibration curves which were obtained with pure reagents or with mixtures of analyzed dolomites, showed a linear relationship with a gradient corresponding to tan ≈ 1. The determination was found not to be sensitive to interference due to a third partner. The relative standard deviation for 0.8% Al_2O_3 was 0.6%.

Glasses and their raw materials [255] were also routinely analyzed. Sample solutions, prepared by decomposition and evaporation with $HF/HClO_4$, were excited in the plasma without further separation. The evaluation was carried out by bracketing the analysis values between standard solutions which had a slightly higher and slightly lower content than the sample.Na, Ca, Mg, Al and Fe could be determined with very high precision. The working of a fully automatic spectrochemical analysis procedure and its routine use were described by Govindaraju et al. [256] for the bulk analysis of silicate rocks and minerals with the aid of a CMP instrument from A.R.L. The decomposition of the sample is carried out by fusing in lithium borate. The fused sample is then dissolved in acid and $Sr(NO_3)_2$ (0.5 g/100 ml) is added to buffer the matrix effects. The sample solutions are automatically analyzed for Si, Al, Fe, Mn, Mg, Ca, Ti, Na and K with a content of a few percent of the element.

Dahmen and Hölzel [247] used the CMP in spectrochemical trace analysis of high purity chemicals; Ca, Cu, F, K, Mg, Na and Pb were determined in NaCl, $NaNO_3$, Na_2SO_4, NH_4Cl, NH_4NO_3, $(NH_4)_2SO_4$ etc. The matrix-containing sample solutions were sprayed into the plasma, and, in addition to the limits of detection of the elements, the stability of the excitation source, the spatial distribution of the trace elements in the plasma and interference were also investigated. Boron and silicon were determined in niobium, molybdenum and tungsten [257]. Dissolution of the sample and the separation of the two elements is carried out in a PTFE vessel. The two elements, after

226

TABLE 5

Some detection limits of the CMP

(Detection limits except for * 2 times the standard deviations of the background fluctuations.

Element	Matrix	Analysis line (nm)	Detection limit (μg/ml)	Plasma gas	Ref.
Al	H_2O	396.2	0.025	Ar	239
As	H_2O	228.8	4	Ar	239
B	H_2O	249.8	0.03	Ar	239
Bi	H_2O	472.3	5	Ar	239
Cd	H_2O	326.1	0.5	Ar	239
Hg	H_2O	253.7	0.04	Ar	239
Mn	H_2O	403.1	0.15	Ar	239
Mo	H_2O	379.8	0.45	Ar	239
Ni	H_2O	361.9	0.3	Ar	239
Sb	H_2O	259.8	0.6	Ar	239
Si	H_2O	251.6	0.2	Ar	239
Te	H_2O	238.6	1	Ar	239
Zn	H_2O	481.1	0.1	Ar	239
Zr	H_2O	339.2	15	Ar	239
Gd	H_2O	432.6	80	Ar	142
Gd	H_2O + 1000 μg/ml Na^+	432.6	2	Ar	142
Nd	H_2O	430.4	250	Ar	142
Nd	H_2O + 1000 μg/ml Na^+	430.4	0.3	Ar	142
Sm	H_2O	442.4	120	Ar	142
Sm	H_2O + 1000 μg/ml Na^+	442.4	0.2	Ar	142
Ag	H_2O	328.1	0.03	N_2	262
B	H_2O	294.8	0.9	N_2	262
Cd	H_2O	228.8	0.45	N_2	262
Co	H_2O	343.4	0.15	N_2	262
Cr	H_2O	360.5	0.06	N_2	262
Cu	H_2O	324.8	0.12	N_2	262
Fe	H_2O	372.0	0.08	N_2	262
Mg	H_2O	285.2	0.016	N_2	262
Mn	H_2O	403.1	0.03	N_2	262
Mo	H_2O	379.8	0.045	N_2	262
Ni	H_2O	441.5	0.09	N_2	262
Zn	H_2O	213.9	1.1	N_2	262
Al	H_2O + 2 mg Cs/ml	396.2	0.02 *	N_2	94
Fe	H_2O + 2 mg Cs/ml	372.0	0.05 *	N_2	94
Mg	H_2O + 2 mg Cs/ml	285.2	0.01 *	N_2	94
Ni	H_2O + 2 mg Cs/ml	352.5	0.02 *	N_2	94

TABLE 5 (continued)

Element	Matrix	Analysis line (nm)	Detection limit (μg/ml)	Plasma gas	Ref.
Pb	H_2O + 2 mg Cs/ml	405.8	0.2 *	N_2	94
Ti	H_2O + 2 mg Cs/ml	498.2	0.15 *	N_2	94
Al		396.1	0.07	N_2	257
As		228.8	3.3	N_2	257
B		249.8	0.02	N_2	257
Bi		306.7	0.3	N_2	257
Cd		228.8	0.13	N_2	257
Co		345.3	0.07	N_2	257
Cr		425.4	0.02	N_2	257
Cu		324.7	0.03	N_2	257
Fe		371.9	0.07	N_2	257
Mo		313.2	0.03	N_2	257
Mn		279.5	0.03	N_2	257
Ni		361.9	0.02	N_2	257
Pb		405.7	0.2	N_2	257
Sb		259.8	0.7	N_2	257
Si		288.1	0.13	N_2	257
Te		238.6	7	N_2	257
Ti		365.3	0.2	N_2	257
V		318.4	0.07	N_2	257
Zr		360.1	2	N_2	257

dissolving the metal at about 65°C in an acid mixture of HF and H_2SO_4 with addition of H_2O_2, are distilled off as BF_3 and SiF_4 at 195°C. The BF_4^- is extracted from the distillate of the HF—H_2SO_4 solution with tetrabutylammonium hydroxide and methylisobutyl ketone [258]. The organic extract is measured in 50 μl quantities with the impulse technique. The limit of detection lies at 0.01 μg/g boron. The calibration curve is linear until 50 μg/g B and at 10 μg/g B the variation coefficient is 2%. To measure Si, the distillate is used directly. Too great a dilution effect can be avoided by throwing away the first few millilitres distilled. Here, the limit of detection is 0.5 μg/ g Si with a linear calibration curve up to 80 μg/g and a variation coefficient of 5% at 20 μg/g Si.

Nakashima et al. described the determination of Fe, Mn and Mg in high purity Mo [259], of Nb, Ti and Zr in steel and other metals [260] and of Ag in biological material [261]. The limit of detection of the CMP for a range of elements has been determined in many publications [50,77,239,240,247,262]. Some of these data are listed in Table 5. A direct comparison of the values with each other does not necessarily give a correct picture of the plasma excitation, since the instrumental details, especially the optical data of the spectrometers used by the individual authors, are very different and affect the limits of detection given. According to Kaiser [263], the smallest measured value $x = x_{Bl} + 3\ s_{Bl}$ is defined. Not all authors hold to this definition, but use twice the root-mean square deviation of fluctuations of the background. These values are given in Table 5. (See also refs. 264—268.)

4. Inductively coupled plasmas

(A) FUNDAMENTAL PRINCIPLES [13,15,19,36,69,105]

The principle of energy transmission in an inductively coupled plasma (ICP) is that of the transformer, in which the primary winding is the HF coil and the secondary the plasma [270,272]. Analogous to the inductive heating of a piece of metal, a fluctuating magnetic field can also couple with a plasma in which ions and electrons are free to move. The HF coil builds an oscillating magnetic field whose force lines run axially inside the coil. If there are charge carriers in this space, they are constrained into circular paths which run concentrically inside the HF coil (eddy current). The electrons, and the ions to a lesser extent because of their relatively high mass, are accelerated through the fluctuating magnetic field. Collisions between electrons and atoms and ions heat the plasma and produce further charge carriers.

In this way, an ellipsoidal plasma can be generated and maintained with, for example, argon, which flows in a quartz tube inside the HF coil. Temperatures of 10,000 K and more are reached. This plasma, however, is not suitable as a spectroscopic light source for various reasons, among them being its high background and low line intensity. In addition, the tube walls are heavily attacked at the high temperatures generated.

In such a plasma, sample injection is a particular problem [19,273,274]. The density of the magnetic force lines decreases rapidly from the outside inwards, as does the magnetic pressure

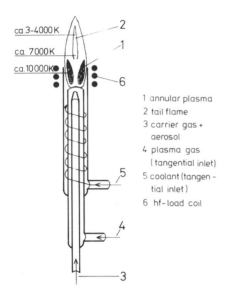

ca. 3-4000K

ca. 7000K

ca. 10 000K

1 annular plasma
2 tail flame
3 carrier gas +
  aerosol
4 plasma gas
  (tangential inlet)
5 coolant (tangen-
  tial inlet)
6 hf-load coil

Fig. 14. ICP torch.

extended by the field on charged particles. Therefore, electrons and ions are forced inwards, thus building up a kinetic pressure. Movement of the atoms perpendicularly to the plasma surface, caused by the thermal heating of the gas, superpose the magnetohydrodynamic effects. These processes offer a resistance to the carrier gas flow and hinder it to a greater or lesser extent from penetrating into the plasma. Hence, the path of the aerosol particles goes preferentially round the outside of this plasma.

If a second quartz tube is arranged concentrically in the first, a cooling gas may be blown in tangentially (see Fig. 14), isolating the plasma from the walls of the tube and at the same time giving the up-to-now elliptical plasma a particular shape in which the lower boundary is flattened. The sample, together with the carrier gas, can now easily be blown into the plasma through a third concentric quartz tube. In this process, a tunnel is formed in the plasma, which takes on an annular, sometimes also described as toroidal, form. In this apparently somewhat complicated form, the plasma source retains all of its desirable properties. The development of this torch, and of the annular plasma, was carried out by Greenfield et al. [61,62,66] at the very beginning of his studies of high-frequency plasma.

230

According to Fassel and Kniseley [273,274] and Boumans et al. [275], the "skin depth effect" plays an important role in the formation of an annular plasma. The eddy currents induced by the magnetic field flow more closely to the outer or skin portions of the plasma at higher frequencies. The skin depth, which is defined as the depth at which the inductive current is $1/e$ ($\sim 37\%$) of the surface value, is inversely proportional to the square root of the frequency, i.e. the higher the frequency, the smaller the skin depth, and therefore the greater the diameter of the tunnel which the carrier gas blows in the plasma, and the easier it is to inject the aerosol and gas into the plasma. According to Greenfield [276], however, the frequency of the HF field has only a small effect on the formation and shape of the tunnel, which, in his opinion, is chiefly dependent on the gas flow and the construction of the torch.

In the upper part of the plasma, the "tail flame" appears (see Fig. 14) as an extension of the tunnel. It is a few cm long, narrow and well-defined. Its course in the tunnel and after is clearly visible. It forms the actual, optically thin spectroscopic light source since it contains, concentrated in a small space, practically the whole sample in the form of free atoms or ions, while their concentration in the surrounding gas is small. Thus line reversal does not exist, or, at high concentrations, exists only within a very small compass. In this way, low limits of detection and linear calibration curves over several orders of magnitude of the quantity of the element injected are obtained.

The plasma core is found inside and slightly above the coil and radiates a very bright, white light. The annular zone is very hot because of the inductive heating, and therefore the spectrum of the neutral argon also shows an intense continuous background from the recombination of ions and free electrons and from the bremsstrahlung. The carrier gas and the sample are only heated inside the channel by radiated and conducted heat from the annular zone. The temperature is therefore lower, of the order of 7000 K, and quickly decreases by a few thousand K in the tail flame. The spectral background also falls decisively with temperature, by about three orders of magnitude in a fall of 10,000 K in the core to 7000 K in the channel. In this greatly reduced background, only bands from OH, CN, C_2, NO and NH of which the OH bands are the strongest, and a weak continuum appear in the region of interest in addition to the Ar lines. Immediately above the core, outside the induction coil, a

TABLE 6

Technical data of some HF generators and ICP torches

HF generator	Type [a]	Power (nominal) (kW)	Frequency (MHz)	Diameter of the torch t[ube]		Middle t[ube]
				Outer tube		
				d_o	d_i	d_o
Radyne S.C. 15	1	2.5	36			
Radyne S.C. 15	1	2.5	36			
Radyne S.C. 15	1	2.5	36			
Radyne R.D. 150	1	15	7			
Lepel T-5-3-MC-J-S		5	3.4	24	22	18
Taylor-Winfield		5	4.8	25		18.5
S.T.E.L. 5060 PL		6	5.4	30	27	21
Philips 131202/01	2	2	50—70		18	
Internat. Plasma Corp.	3	2	27.12	20	18	16
Société Francaise S.T.E.L.		6.6	5.4	30	27	22
A.R.L.	3	2.6	27.12			
Kontron/Linn FS 4	2	4.0	27.12	20	17.2	15.8

[a] 1, Free running; 2, with capacitive tuning; 3, crystal-controlled.
[b] N_2.

further plasma zone forms. Here again, the plasma is very bright, but now appears more transparent. Inside this zone the intensity of the background continuum falls very sharply and it is here that the highest signal-to-noise ratios are usually observed.

The residence time of the atoms in the plasma, of the order of a few msec [273,274], is relatively long and is, in the main, determined by the speed of the carrier gas. This length of time in the plasma and its high temperature are sufficient to atomize the sample almost completely. On the other hand, the time is short enough to prevent the reformation of thermally stable compounds. Because the processes happen in an atmosphere composed of a noble gas, the working gas cannot take part in any chemical reaction, which is often a source of interference in combustion flames.

The gas temperature in the tail flame is dependent on the heating time. Since this, in its turn, is dependent both on the flow rate and the length of the tunnel in the annular zone, and the size of the plasma and the tunnel are dependent on the power, there is a direct

(mm) Inner tube		Gas flows (l/min)			Turns of the work- ing coil	Ref.
d_0	d_i	Coolant	Plasma gas	Carrier gas		
		17	5		3.5	61,62
0.6	0.5	18	3	0.5	3	329
		7.5	7.5	0.5	2.5	345
	2.0	20—70 [b]	10—35	2—3		68,272
7	5	22	0.4	0.5	5	71
		30	1.7		2	271
	1.5	22	7	>1.5	3	269
7	1—1.5	15		1—2	2	303
	1.5	10		1.4	2	318
8	2	20	7	2	3	369
	1.5	10.5	0.5	1.0		
6	2	15		1	2.5	

link between power and temperature. For an increase in temperature, the power must be increased more than proportionally. The frequency has little or no effect on the temperature.

According to Barnes and Schleicher [277], high frequencies above 20 MHz are preferred since small plasmas can be heated more efficiently at higher frequencies and low power operation is possible. Also, the power is then concentrated more in the outer regions of the plasma, which is more easily penetrated, and is less likely to be extinguished by a high velocity central gas stream (but see refs. 36, 63, 276). At higher frequency and constant power, the temperatures are somewhat lower, so that the continuous background is reduced.

Since, in addition to the high temperature and relatively long residence time, about 2.5 ms [278], of the atoms in the tail flame, there is also a chemically inert atmosphere, the prerequisites for a high degree of atomization are fulfilled. If the sensitivity of a spectral line, which is defined as the slope of the calibration curve, is recorded against temperature, a maximum is obtained at the so-called norm temperature [19,272], which, however, lies below 10,000 K for only a few elements (see Table 1, p. 176). The curve gives the population of the corresponding excitation state of the atoms (Boltzmann distribution). With increasing temperature, more and more atoms are raised to this energy level, but from a certain temperature, the population decreases again since excited atoms pass on to the next energy state.

Since a certain temperature distribution is found along the axis of the tail flame, the individual elements have the maximum intensity of their emission lines in different zones, that is, for every element there is a zone of maximum sensitivity. This means that, in the simultaneous analysis of many elements, a compromise must be reached for the observation level. And as the temperature zones also vary with altering plasma power, at each change of power the observation level must be determined again. Gas and excitation temperatures and the temperature distribution vary with parameters such as size and construction of the ICP burner used, as well as with power and frequency of the generators [36,269,277—300].

A rough summary of the details may be given as follows: local thermodynamic equilibrium (LTE) cannot be assumed to exist [301]. The maximum temperatures, depending on the conditions, lie between 5400 K [293] and 10,000 K [299]. The radial temperature distribution in the plasma shows a maximum that lies off the axis. This maximum may be more or less strongly pronounced and seems also to depend on the injected sample. The sample causes a considerable drop in temperature at the centre and a constriction of the plasma, by which the temperature is raised [293]. The temperatures in the tail flame lie a few thousand K lower, but even so are still higher than those of combustion flames by a factor of two or three.

The dimensions of the annular plasma are determined by the working frequency. At low frequencies, the skin depth is large and the power absorbed by the plasma is distributed over a large volume [272]. A higher powered generator is then necessary to maintain the same excitation conditions and a stable plasma. At higher frequencies, correspondingly lower powers are needed. The higher the power, however, the higher the gas temperatures that can be reached. In this way, chemical interference is less [see Sect. 4.C(1)]. The plasma can take up larger quantities of sample material and solvent, and so the desolvation of the aerosol [see Sect. 4.B(2)] and its attendant problems may be avoided. An increase in precision is often noticed

[36]. High powers also allow the use of diatomic gases as coolant.

If one compares the data given in the literature on the properties, efficiency, and interference that occurs in inductively coupled plasmas, one notices many contradictions. The situation may be clarified if it is remembered that the ICP cannot be spoken of as one uniform excitation source, but each torch is an individual instrument. Most of the instruments so far described were constructed by the individuals concerned and so they differ to a greater or lesser degree in their dimensions, data of the HF generators, and in many other particulars (see Table 6).

(B) INSTRUMENTATION

Figure 15 shows schematically the construction of an ICP burner with generator and spectrometer. The actual burner, 1, consists of three concentric tubes of silica or glass (see also Fig. 14). These tubes may be fused together or be connected through joints to enable them to be cleaned better. The optimal operation of the torch depends on the correct and precise fitting of these tubes and particularly on the correct positioning of the opening of the inner tube, through which the aerosol is to be blown into the plasma, in relation to the HF coil. In particular, during ignition of the plasma, tubes which are incorrectly adjusted can melt because of overheating in a very short time.

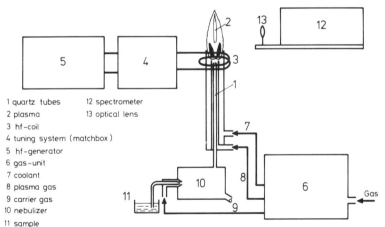

1 quartz tubes   12 spectrometer
2 plasma   13 optical lens
3 hf-coil
4 tuning system (matchbox)
5 hf-generator
6 gas-unit
7 coolant
8 plasma gas
9 carrier gas
10 nebulizer
11 sample

Fig. 15. Schematic diagram of an ICP.

Through these tubes flow three different gases: the coolant gas, the plasma gas and the carrier gas (see Fig. 14). Either Ar or N_2 is used as coolant. Ar is a poorer conductor of heat than N_2 and so protects the surrounding silica glass from overheating better, but it is expensive. The spectral background is lower with Ar than with N_2. Greenfield [19], however, used N_2 because with this coolant better signal-to-noise ratios, and therefore lower powers of detection, were obtained for a number of elements. These S/N ratios are, however, dependent on the HF power. N_2 is preferred at a power of 3 kW, and also 6 kW for elements with high excitation energy, and Ar at 6 kW for elements with low excitation energy. N_2 is also preferred for high power ICPs, which need a high coolant gas flow [19] (see also refs. 280, 281, 302). The protecting gas flows at a rate of about 10—20 l/min. In the literature, values between 7 and 70 l/min are given for this gas flow, depending on the construction.

Although N_2 or He can also be used as the plasma gas, the ICPs described in the literature work, almost without exception, with Ar since it has various advantages. Being a noble gas it undergoes no chemical reactions, has a low spectral background, reaches a higher temperature for the same power, and can easily be ignited (see Table 6). Typical flow rates of the plasma gas are in the range 1—7 l/min, but higher flows, up to 35 l/min [272], are also used. Some authors work without the plasma gas at low total gas consumption [12,303,304].

The sample reaches the plasma with the carrier gas through the innermost of the three concentric tubes, the injector. Here, also, Ar is used. The gas stream, with flow rates between 0.5 and 3 l/min, must be very finely adjustable since it determines the time the sample remains in the plasma and, therefore, the temperature of the tail flame. According to Greenfield et al. [272], the opening of the injector is also critical and co-determines the properties of the torch. Small deviations in the dimensions of the other parts can be corrected with the gas flow for the plasma gas and coolant.

Regulation and constancy of the three gas streams are provided for by a flow meter, pressure regulator, and, in the aerosol flow, a flow regulator.

The burner system is kept adjustable by a system which allows each desired zone of the plasma, which is observed with the spectrometer, to be adjusted accurately and reproducibly both laterally and vertically. Since the position of the HF field in relation to the burner must be defined exactly, the coil is also fixed by this support. Burner, support coil and, if present, the matching system are placed in a metal housing, the purpose of which is to screen the HF field. By this, not only are the requirements of the law fulfilled, but the sensitive measuring devices and the electronics of the spectrometer are also protected from the HF radiation. Improper shielding can also cause instability in the operation. To observe the plasma, there is an opening in the wall of the housing. An additional window of dark blue glass enables one to observe the plasma with the naked eye without harm from UV light.

(1) HF generators

A HF (or RF) generator [29,305] produces the high frequency alternating current at high power which flows in the working coil at up to 100 A and whose frequency for an ICP, can be between about 1 and 100 MHz. For the oscillator, certain modifications of triode circuits, as shown in Fig. 16, or transistorised circuits with a final valve stage are possible. The oscillating circuit LC determines the frequency; the oscillator operates by feedback to the triode grid.

The HF power can be coupled [306] to the plasma either over the output coil of the tuning circuit [Fig. 16(a)] or directly, if the working coil is part of the generator circuit. The feedback with the so-called Colpitts circuit [Fig. 16(b)] is achieved by dividing the capacitor in the circuit into two individual condensers, whose ratio determines the feedback factor. In the Huth—Kühn circuit [Fig. 16(c)], the small capacitance between grid and anode of the triode acts as the feedback route. The two oscillating circuits L_1C_1 and L_2C_2 are tuned to nearly the same frequency. Figure 16(a) shows the circuit diagram of a typical HF generator as used for ICPs. The output coil, 3, and the variable capacitor, 4, form, together with the working coil, 5, a secondary oscillating circuit, which can be tuned to the working frequency. The coupling with the primary circuit is inductive. The mismatching of the circuit caused by the introduction of the aerosol into the plasma can easily be compensated for using this circuit.

Three types of oscillators are differentiated [36].

(a) A free running oscillator, in which the oscillator coil is also the working coil for ICP.

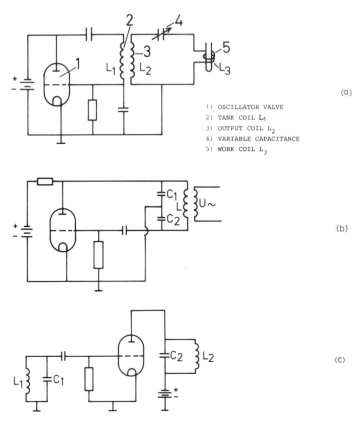

Fig. 16. HF oscillator circuits (a) for a typical HF generator for use with ICP; (b) basic circuit of Colpitt's HF generator; (c) basic circuit of a Huth—Kühn generator.

(b) An oscillator with capacitive tuning.

(c) A crystal-controlled oscillator.

In the free running oscillator, the frequency is determined by the capacitances and inductances in the oscillating circuit, i.e. partly by the working coil, whose inductance depends on the impedance of the plasma and degree of coupling between coil and plasma, as well as on the number of turns of the coil and its diameter. That means that, with every change in the aerosol that is sprayed in, the frequency can alter slightly. The power absorbed by the plasma, on the other hand, remains constant. For small changes in the impedance of

238

the plasma, a free running generator is self-compensating. Tuning is not possible in this system. It is possible with the second type, where the working coil is in a second oscillating circuit which is inductively coupled with the tank circuit [see Fig. 16(a)]. Its resonant frequency can be altered by a variable capacitor. Frequency changes due to alterations in the plasma impedance are, here again, slight. Tuning is carried out either by hand or automatically [275] and is required mainly during the ignition of the plasma.

In the third kind of oscillator, the frequency is controlled by an oscillating crystal and kept constant. This kind of generator is more difficult to build since, in addition to the crystal controlled oscillator, a power amplifier is needed to produce the high output power. Frequency changes are not possible, so a change in the plasma impedance can lead to heavy losses of power and if there is strong interference the plasma may be quenched [74]. Crystal controlled units therefore possess a tuning system, again in the output circuit, with which frequency changes can be smoothed out, partly automatically. With such a generator, the ignition process is relatively long-winded since it is at this moment that the greatest impedance changes appear. Neither of the other two kinds have any such difficulties, or only to a considerably lesser extent.

Due to their construction, the three kinds of generator have different advantages and disadvantages. The most important requirement in the operation of an ICP is that constant power is absorbed by the plasma. Whether the stability of the frequency is an equally important factor cannot be decided at present with any certainty [307] as the views of different authors on this point are not in agreement. According to Greenfield et al. [36], no complicated tuning devices are necessary since important impedance and frequency changes are not observed when the aerosol enters the plasma [272] (maximum 1 kHz above and 1.5 kHz below). The calibration curves, which are linear over several orders of magnitude, also indicate no noticeable changes in the electrical conductivity of the plasma.

A further advantage of free-running oscillators and those with capacitive tuning has yet to be mentioned: they are of simpler construction and therefore cheaper.

The working coil, whose dimensions are fixed, since they determine the frequency of the generator tuning circuits, consists of about 2—5 turns of a copper tube of a few mm diameter. The coil itself has a height of 10—20 mm and an inner diameter between 20 and

40 mm. The latter must be adjusted to the burner diameter so that the distance between coil and tube is only a few mm. The high-frequency alternating current flowing in the coil can be up to some 100 A. Cold water flows through the coil to remove the heat generated by the current.

A few additional arrangements in the generators shall also be mentioned. Various instruments are equipped with bidirectional couplers, which measure the forward and reflected power simultaneously. On other measuring instruments, the anode current and output power of the transmitter valve may be shown.

Filters applied between generator and matchbox prevent the emission of unwanted upper harmonics. Voltage control ensures that variations in the circuit do not show up as variations in the output power. Automatic power regulation in the region of 1—3% of the operating level, in order to achieve sufficient stability of the plasma emission, and a protection circuitry which limits the forward and reflected power in order to prevent damage to the generator, are important components in an ICP system.

Various opinions are held about the significance of the frequency level for the formation of the annular plasma [36,64,273—275]. Abdallah et al. [308] compared generators at 5 and 40 MHz equipped with an identical device for introducing the sample into the plasma and similar spectrometers. At 40 MHz, the temperatures and the electron density are somewhat lower than at 5 MHz, but with the 40 MHz ICP, the authors obtained better detection limits because of significant decrease in the background emission.

On the other hand, it is certainly much easier and cheaper to build generators with high power if the frequency is low. Such generators with powers up to 15 kW and a frequency of 5—10 MHz are preferred by Greenfield et al. [36,272,276] who thus obtained higher sensitivity and somewhat better precision and a lower spectral background (band spectra). Other teams prefer HF generators with higher frequencies and power between 1 and 5 kW [275,303,309—318].

The design of the ICP ensemble depends on the available power, e.g. the nebulizer system, desolvation, number of gas flows and the type of cooling gas. At lower powers, under 2 kW, two gas flows are sufficient for the burner, namely carrier gas and coolant. For higher quantities of solvent, desolvation is neccesary. If organic solutions have to be analyzed, normally three gas flows are necessary to separate the plasma from the injector tube.

Higher powers make desolvation superfluous and N_2 may be used as coolant as well as Ar. In addition, no problems with drawing air through the nebulizer are observed. Large amounts of gas are, however, needed.

The performance data mentioned here are the nominal powers given by the generator, of which only a part reaches the plasma; generally, the effective power is about 60%. Table 6 gives some data on generators and burner types with operating parameters which are described in the literature.

(2) Sample introduction

The sample solutions are made into aerosols by a nebulizer system and transported with the carrier gas into the plasma [244,319—323]. There are two possible methods of nebulizing the sample, the pneumatic system and ultrasonic nebulization [324]. After the aerosol is formed, water or organic solvents can be removed from the gas with a desolvation unit [325] if the ICP cannot take larger quantities without undergoing sensitive changes. Whether desolvation is necessary or not depends on the construction and the working parameters [36]. It is particularly necessary at low HF powers.

The construction of a pneumatic nebulizer with a spray chamber to remove droplets is the same as those used in AAS, and does not need to be described in any further detail.

A problem is often presented by the quantity of gas necessary for the nebulizer. Argon is used almost without exception at flow rates between 0.5 and 3 l/min. The flow rates of the carrier gas, especially the lower rates, are not sufficient for optimal aerosol formation with all types of nebulizer. It is this gas flow that, in the main, determines the analytical properties of the plasma (see above) and, therefore, its rate must be adjusted to suit the burner, not the nebulizer. This means that, in some cases, only particular kinds of nebulizer can be used [244—246,327] to introduce a sufficient amount of sample with low gas velocity constantly into the discharge. A nebulizer with a very low gas consumption is described by Meinhard [245]. Four other types, the cross flow, the threaded cross flow, the Babington, and the fritted disc nebulizer, were investigated by Apel et al. [326]. With all types, even small fluctuations of the gas flow cause consierable changes in the emission line intensity and thus the flow rate must be carefully controlled [328].

In addition to the indirect nebulizers, in which the aerosol is produced in a separate chamber and then transported into the plasma with the carrier gas, there are also direct nebulizers, in which the injector of the torch consists of a pneumatic nebulizer. These types are chiefly used for organic solutions and since they are difficult to build, are only seldom used.

Figure 17 shows schematically an ultrasonic nebulizer with a desolvation unit attached. The sample solution is admitted into the aerosol chamber, 8, with a pipette, peristaltic pump or an automatic dosing system. The water, 3, couples the ultrasonic waves from the transducer, 1, to the sample solution, 5, whose surface breaks up and produces the aerosol. For better coupling, only a very thin nylon or polythene foil, 4, separates the water and sample chamber. In other types there is no ultrasonic focussing by water and the sample solution is injected onto the surface of the ultrasonic oscillator.

The aerosol is carried out by the argon and heated to 200—500°C in the furnace, 9. It then passes through a cooler, 10, where the evaporated solvent condenses and leaves the system by means of the drain, 11. The remaining dry aerosol particles reach the plasma. The aerosols obtained by ultrasonic means are finer and more homogeneous than those produced by the pneumatic nebulizers.

The quantity of solution nebulized per unit time is also greater, and this is often noticeable in better powers of detection [311]. Since the gas flow does not take part in the nebulization process, the nebulizer can work with whatever quantity of gas is required for the plasma. Stable operation of the nebulizer is also important here,

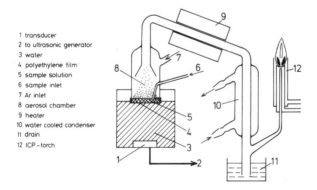

1 transducer
2 to ultrasonic generator
3 water
4 polyethylene film
5 sample solution
6 sample inlet
7 Ar inlet
8 aerosol chamber
9 heater
10 water cooled condenser
11 drain
12 ICP-torch

Fig. 17. ICP torch with an ultrasonic nebulizer and a desolvation unit.

since the stability influences both the line and background intensity and, therefore, also the noise level and the detection limits. Working with the ultrasonic nebulizer is, however, a lengthy process. Unused solution must be removed before the introduction of the next sample and the whole chamber must be washed out several times with distilled water or the next sample solution. To overcome these problems, Hoare and Mostyn [329] developed a multicell sample vessel, with which the plasma discharge does not have to be interrupted while the sample is being changed.

In the combination of a nebulizer of high efficiency, such as the ultrasonic nebulizer, and a low power plasma, desolvation is necessary. With a low power ICP, the free molecular gases produced by the solvent provide interference [74] since they alter the impedance and use up energy for their dissociation and excitation. Without desolvation, a higher power would be needed to retain the same powers of detection. So desolvation here contributes to the low limits of detection [330]. However, new sources of interference appear. Because of the relatively large volume and surface area, a memory effect makes itself noticeable, by which changing the sample is made even more lengthy. Matrix influences in connection with the desolvation have also been observed [311,331]. So, for example, the intensity of the emission lines of some elements can greatly depend on the temperature of the desolvation chamber. Besides the usual nebulizers, other sample injection systems are also described in the literature [332]. In general, all of the procedures described in Sect. 3.B(1) for MIPs can be used for ICPs, e.g. heated graphite atomizer [333], hydride generation, or reduction to the metal as for Hg [334]. Greenfield and Smith [68] determined trace elements in μl samples, which were injected with a μl pipette into the pneumatic nebulizer. This procedure is similar to the impulse technique [248], which is described in connection with the CMP [see Sect. 3.E(1)]. Kniseley et al. [335] determined μg/ml traces in 25 μl samples in biological fluids with a similar technique. Nixon et al. [336] used a tantalum filament vaporization system as a sample introduction device with a maximum sample volume of 200 μl. The sample is placed by a syringe on a Ta strip (37 X 10 mm) held under a silica cover, the solution is evaporated by ohmic heating of the Ta strip, and the residue is then evaporated at about 1800°C. With the Ar carrier stream (1.2 l/min) the vaporized sample is fed into the plasma. The limits of detection lie in the ng/g range or lower. Since the authors injected the sample solution through a septum, no air could penetrate into the plasma. Kleinman and Swoboda [337] evaporated the sample from a graphite disc which was also heated by ohmic current. Human et al. [338] used a spark as a sampling and nebulizing device. The solid sample is vaporized by a spark discharge (12 kV) in a chamber through which Ar flows. Fused metals such as tin base solder were nebulized by Fassel and Dickinson [72] by ultrasonic energy. The ultrasonic probe was brought into contact with the molten metal surface. The resultant aerosol quickly solidified in the argon stream which carried it to the plasma.

Various experiments have been tried to bring solid, pulverized samples into the plasma [339,340]. Reed [341—343] originally developed this technique to grow crystals with the ICP. However, for analytical purposes there is the great problem of pulverizing the sample to produce a constant, reproducible powder, then injecting it into the plasma, where the size of the grains should be more or less the same. In a review, Greenfield et al. [36] established that, so far no satisfactory solution has been found for this problem. Hoare and Mostyn [329] tried to produce a powder aerosol with a shaker in an argon stream blowing over the vibrating sample. The capillary tube of the plasma burner reaches down into the sample chamber, directly over the powder. The authors report that the excitation of poweder in the plasma is not very effective. If the particles are too big, they are not completely vaporized. Low injection velocity and small particle size are essential. The spectral emission is also different from that of solutions.

Pforr and Aribot [344] also tried to inject solids or viscous fluids into the plasma. They fed solids directly into the plasma, but could detect Fe, Mg, Pb and As only at concentrations over 1% in glass and silica rocks. For powders, they used a similar system to Hoare and Mostyn's [329] and could reach the μg/g region. Dagnall et al. [345] used a fluidized bed chamber. The powder lies on a porous glass disc in a glass tube. Argon flows through the powder from underneath and suspends it. Mechanical shaking aids the process. For impurities of Be and B in MgO with a concentration of 10 μg/g, they reached relative standard deviations of 6.5 and 10%.

(3) Instruments

The importance of the ICP as a spectrochemical excitation source is shown by the number of commercially available instruments,

244

whether as a complete unit with a spectrometer as a modular system, or for adaptation to an existing spectrometer. The following list merely serves as a short review and makes no attempt to be complete.

A very variable instrument is offered by the firm Kontron, Eching (München), West Germany, with their ICP Plasmaspec [346]. The instrument consists of various parts; the generator with coupling unit (either with inductive adaptation 2.5, 4 or 7 kW at 27.12 MHz in the Huth—Kühn circuitry, or with a crystal controlled oscillator), the source unit with pneumatic atomizer, atomizer chamber, torch assembly and gas supply, and the "Sequential Photometer Model HRS2" with holographic grating and readout system Model LEP 500. If desired, a multi-element spectrometer model Z100 with 1 m focal length and up to 30 channels is available.

Applied Research Laboratories (ARL), Sunland, California, U.S.A., offer an ICP instrument in their "Automatic Liquids Analyser" 33000LA with crystal-controlled HF generator (27 MHz at 3 kW). Besides the standard model with a sequential spectrometer, an ICP with a monochromator and manual wavelength adjustment or a multi-channel spectrometer is available.

The "Plasma/AtomComp. Direct Reading Spectrometer System" from the Jarrel-Ash Division of Fisher Scientific Company, Waltham, Mass., U.S.A., consists of a complete ICP excitation source with generator and an 0.75 m spectrometer, that is controlled from a PDP 8 control processing unit.

The ICP-Spectrometer Model ICP 2100 from Labtest Equipment Co., Los Angeles, California, U.S.A., combines a Plasmaspec-ICP source, with crystal-controlled generator (27.12 MHz, 2 kW) and a 1 m Paschen—Runge spectrometer with maximum 30 photo-multiplier tube detectors and a Readout Data Acquisition and Display System Model CRT-1000.

In addition to the complete ICP spectrometers, there are also ICP discharge units on the market such as the RF-Plasma Emission System from Plasma-Therm., Inc., Kresson, N.J., which consists of the HF generator (27.12 MHz, 2.5 kW) with a semiautomatic tuning device, an automatic power control and the ICP torch. Further HF generators are listed in Table 6. (See also refs. 347—354.)

As in all spectroscopic procedures, the prerequisite for analytical work with the ICP is the optimal adjustment of the radiation path. If the generator and the electronics of the measuring instruments are warmed beforehand, the plasma can be ignited [66,72,318,345,355]. This does not happen of itself when the HF current is switched on, since the argon in its normal non-conducting state cannot couple with the electromagnetic field. In many cases ignition of the plasma is a somewhat critical process, since during it the impedance of the coil alters considerably. The operating instructions must be followed exactly, especially with crystal-controlled generators, to avoid the high power losses occasioned by even small mismatching. It is also at ignition that the burner tube system can be overheated and damaged by incorrect procedures. Useful aids at this stage are instruments measuring the absorbed and reflected powers, which makes for quicker tuning.

The burner tube system is first flushed out with argon to remove air which would otherwise be a source of interference. While the carrier gas is switched off, coolant and plasma gas are set to low flow rates. In some instruments, the coolant is not switched on until after ignition. After the HF power for the working coil is set low and the system is pretuned, the plasma is initiated with seed electrons. This can be done in various ways. The simplest way is by producing free electrons and ions in argon with a Tesla coil, a high voltage spark, or with high tension discharges which leap across the silica burner tube near the coil. Another possibility is introducing an unearthed carbon rod into the burner from the open end to the region of the injector tip. In the HF field, the graphite rod heats up and finally emits free electrons which start the plasma. Once the plasma is going, the rod is then slowly removed from the burner. Eckert [270] describes a further method, in which a tungsten wire is introduced into the burner tube to the region of the coil end. Then a capacitive current flows through the tube dielectric between coil and wire. If the wire is pulled into the centre of the burner, a filamentary arc appears which finally ignites the plasma. During the ignition procedure, it is important to readjust the tuning of the system. Once the plasma is burning, either the output power of the generator must be immediately lowered or the coolant flow increased to protect the burner from overheating. Then the three gas streams and the power can be set to the working values and a last adjustment follows. Allemand [356] describes the initiation of the induction-heated plasma at low power levels from 250 to 750 W as a two-step process, in which a filamentary plasma with high electrical resistance is first formed, while in the second step the power is raised, the plasma fully ignites and exhibits low resistance. With a slight modification of the matching circuit, the authors are able to initiate the plasma automatically without tuning components. When the plasma is burning stably, after about 15—30 min, the image of the plasma zone which is best for the analytical problem is formed on the entry slit of the spectrometer or on the grating, after which everything is ready for analysis. The plasma should burn during the entire working time, and should not be put out during short pauses.

Usually, there is a large enough choice of lines for analysis to cover the required concentration range or to avoid spectral interference. Both atomic and ion lines can be used. With the most sensitive spectral lines, calibration curves with a linear range over 2—5 orders of magnitude can often be obtained [357] so that they can be used both for trace analysis and for the determination of the chief components.

The limits of detection reached with this excitation source lie in the ng/g range and lower μg/g range. Since the calibration curves are linear over several orders of magnitude, calibration is usually without problems. All the other usual calibration procedures, such as standard addition or bracketing the analysis values with standard solutions, are also possible. Because of the lesser interference that this excitation source is subject to, it is in many cases possible to use pure aqueous standard solutions for calibration without considering the concentrations of other elements in the sample.

Since the spectral background can change with the composition of the sample, a background correction is advantageous to ensure that a serious error is not made.

The optimal parameters for the analytical problem, such as power, coolant gas flow, plasma gas flow and carrier gas flow, quantity of solution nebulized, and observed plasma zone, must be ascertained for each individual case. In particular, careful measurements must be made for establishing the best plasma zone since each element shows a maximum intensity of its spectral lines at a particular height. In simultaneous multi-element analysis, a compromise must be reached for the observed plasma zone. In this, other important operation parameters, such as the power coupled to the plasma and the flow rate of the carrier gas, must be taken into consideration [310,311,358]. As an example, the compromise conditions established by Boumans and de Boer [310] for a low power ICP are power of 0.7 kW, carrier gas flow 1.3 l/min of Ar, observation height 15 mm above the top of the coil.

(1) Interference

Although the ICP is not totally free of interference, it can be considered in OES to be the excitation source for solutions which, in comparison with the others, shows the fewest possibilities for error [359,360]. Contradictions about interferences in ICP—OES which come up frequently in the literature on this subject stem from the fact that there is no standard ICP torch but many different variations with individual properties.

Sources of interference can be divided into four groups [331,361].

(a) Nebulization interference. These are caused by the properties of the analysis solution. Changes in the analysis solution (excepting changes in the concentration of the analyte) influence the number of atoms of the analyte element in the aerosol.

(b) Chemical interference. Here the presence of an element or a compound leads to a lessening of the number of free analyte atoms in the plasma.

(c) Ionization interference. These appear when the electron density in the plasma is increased due to the presence of an easily ionizable element, and the ionization equilibrium of the analyte is altered.

(d) Spectral interference. The analysis line overlaps other spectral or band lines in this form of interference.

(a) Nebulization interference. Interference due to spraying the sample solution are chiefly noticeable if the quantity of the sample nebulized per unit time varies considerably. This happens in the presence of foreign salts or organic compounds and solvents by alterations in the viscosity, the surface tension or the density of the solution. This occurs to some extent even for solutions with high concentrations of mineral acids [361,362]. Apart from this, memory effects may result if there are long tubes and large vessel surfaces in the nebulizer and desolvation system. Interference of this sort is however, not the fault of the plasma source itself and can be easily avoided and controlled.

248

Interferences caused by desolvation have been described by Boumans and de Boer [311]. They are related to the difference in volatility between matrix and analyte, and are noticed as a change of concentration of the analyte in the aerosol during desolvation. So, for example, interference by KCl and other salts in the determination of Fe, Zn and Cd, which the authors ascribe to desolvation, have been noted. The mechanism of these and similar effects is at present not understood.

Kirkbright et al. [363], during the determination of iodide, observed interference by Cu which they explained as follows. In the cloud chamber of the nebulizer, a reaction proceeds between I^- and Cu^{2+}

$$2 \; Cu^{2+} + 4 \; I^- = 2 \; CuI + I_2$$

In this way, the concentration of free I_2 in the gas increases, while in the drain solution that leaves the aerosol, it decreases. Iron(III) and other oxidising agents show the same effect.

Difficulties of quite a different nature can occur in the analysis of organic samples, such as oils or organic solvents, if the injector tube orifice clogs with carbon deposit after prolonged spraying.

(b) Chemical interference. Due to the very high gas temperature, long residence times, and the inert atmosphere in the plasma [12], chemical interference caused by the formation of thermally stable compounds or radicals is very small, indeed generally negligible. The "classical" interference in the determination of Ca due to PO_4^{3-} and Al, investigated by many authors, serves as an illustration. It appears in combustion flames, d.c. plasmas, CMPs and MIPs, but not in ICPs [61,62,74,273,274,311,331,364—366]. Veillan and Margoshes [271] observed an enhancement effect due to PO_4^{3-} on Ca, instead of a reduction, and a mutual interference between Cu and Al. Only Kornblum and de Galan [293] report serious interference effects from PO_4^{3-} on Ca with a low power ICP. In short, one can say that the combustion of the solution, as regards accompanying elements, anions, or changes in the matrix compositions, has little or no effect on the intensity of the spectra lines of the analyte. In all cases, operating conditions may be so chosen that any interference affects the analytical results by only about 10%. With increased plasma power, this chemical interference becomes even less.

(c) Ionization interference. Even easily ionizable elements, which can lead to serious interference in MIPs and CMPs, cause little or no change in the intensity of the spectral lines in ICPs [365,367]. According to Boumans and de Boer [12], this sort of interelement effect is, however, the most important form of interference in ICP — excitation. It can be largely eliminated by the use of spectroscopic buffers. As an example of this kind of interference, the rise in the intensity of the lines P 253.5 nm and Cr 425.4 nm in the presence of 100 μg Na/ml by a factor of 2.4, as observed by Mermet and Robin [364], may be mentioned. It was established that, at concentrations up to 1 mg Na/ml, the temperatures and electron density are not particularly influenced [331]. Large quantities of easily ionizable elements do not seem to affect the impedance of the plasma either, since the frequency, as well as the reflected power, remains almost constant.

The reports about observed interference are, as mentioned above, partly contradictory on this point too, as they are, for example, about inexplicable changes in the background in some regions of the spectrum when certain elements are present [330], about changes in the spatial distribution of radiating species in the plasma when the matrix is changed, and about the influence of Na [12,241,273,274,331,364,367].

These effects depend on the experimental parameters such as observation height, power output to the plasma, flow rate of the carrier gas, type of nebulizer and temperature of the desolvation apparatus [311,323,330,365]. The interference generally becomes less with increased power, decreasing observation height, and decreasing carrier gas flow. The theoretical connection is at present not known [12].

(d) Spectral interference. As in other excitation sources, line overlapping is observed in the ICP. Since, however, on one hand the spectra are relatively simple over the hot central core [368] and on the other hand because of the high excitation temperatures, lines with higher excitation potentials also appear in the spectrum along with the ground lines, it is seldom that interference of this kind cannot be avoided. One can usually use other sufficiently intense lines.

For all the sources of interference discussed here, it may be said that the operating conditions of the ICP can be adjusted so that interference effects can be overcome [311]. Relative to other excitation sources, they are small. According to Greenfield et al. [272], the higher power causes, in addition to better limits of detection and

250

precision, an elimination of chemical interference and of band spectra.

(D) PRACTICAL APPLICATION

Its excellent properties as an excitation source for optical emission spectroscopic analysis and its low susceptibility to interference have helped the ICP to spread rapidly and gain universal application within a few years of its first use by Greenfield in 1964. Although this plasma with its high powers of detection appears especially suitable for trace analysis, main and secondary constituents have also been determined with it. Since matrix effects are small, time-consuming and long-winded separation operations are done away with and the sample solutions are injected directly after decomposition. Of the great range of literature on the subject, only a small and necessarily arbitrary selection can be discussed here (see also Table 7).

Souillart and Robin [369] worked with a generator at 5.4 MHz and 6.6 kW and determined the limits of detection, which lay in the region of 0.4—0.0005 μg/ml, for 27 elements in pure aqueous solution. They also investigated the influence of up to 5 mg Zr/ml on the determination of Hf in the μg/ml range, in which they established depressions at Zr concentrations above 50 μg/ml. Butler et al. [368] compared two different excitation sources combined with three different spectrometers in the determination of alloying and impurity elements in low and high alloy steels. They operated the first ICP with an HF generator from Lepel High Frequency Laboratories model T-2, 5-1-MC 2-J-B at 29 MHz and 2.5 kW. The aerosol, produced by a pneumatic nebulizer, was desolvated. The radiation from the source was examined simultaneously in two spectrometers: (a) a Hilger Engis model 1000 1m-Czerny-Turner mounting scanning spectrometer and (b) an Applied Research Laboratories (FICA-France) Quantoscan 35 1m-Czerny-Turner mounting scanning spectrometer. The second ICP worked with a plasma power supply from Lepel High Frequency Laboratories model T-5-3-DF1-2-J-S at about 30 MHz and 5 kW. Here the pneumatic nebulizer was used without desolvation apparatus. A multichannel direct reading 1m-Paschen-Runge mounting spectrometer model QVAC 127 was used from ARL. The samples (1 g) were dissolved in a mixture of 30 ml HCl (1 : 1) and 5 ml HNO_3 and then diluted until the solution contained 5 mg sample in 1 ml. Synthetic reference solutions were

TABLE 7

Some detection limits for ICP

Element	Analysis line (nm)	Matrix	Detection limit	HF generator		Nebulizer
				Power (kW)	Frequency (MHz)	
Al	396.2	H_2O	0.2			
As	228.8	H_2O	6			
B	249.8	H_2O	0.1			
Be	234.9	H_2O	0.003			
Ca	393.4	H_2O	0.0001			
Cd	228.8	H_2O	0.2			
Ce	418.7	H_2O	0.4			
Cr	357.9	H_2O	0.1			
Cu	327.4	H_2O	0.06			
Fe	259.9	H_2O	0.09			
Ga	417.2	H_2O	0.6			
Ge	265.1	H_2O	0.5	0.7	50	Ultrasonic
La	408.7	H_2O	0.1			
Mg	279.6	H_2O	0.003			
Mn	257.6	H_2O	0.02			
Mo	317.0	H_2O	0.5			
Nb	309.4	H_2O	0.2			
Ni	352.5	H_2O	0.2			
Pb	283.3	H_2O	2			
Sn	303.4	H_2O	3			
Ti	334.9	H_2O	0.03			
W	276.4	H_2O	0.8			
Zn	213.9	H_2O	0.1			
Zr	343.8	H_2O	0.06			
Al	396.2	H_2O	30			
Au	267.6	H_2O	40			
B	249.8	H_2O	30			
Mg	279.6	H_2O	0.05			
Hf	264.1	H_2O	40			
Nb	318.0	H_2O	2	6.6	4.5	Ultrasonic
Ta	296.3	H_2O	30			
Ti	334.9	H_2O	2			
V	309.3	H_2O	1			
W	400.9	H_2O	100			
Zn	334.5	H_2O	50			
Zr	339.2	H_2O	5			
Al	396.2	H_2O	1			
Ba	455.4	H_2O	0.1			
Ca	393.4	H_2O	0.5	1	27	Pneumatic
Co	345.4	H_2O	4			
Cr	425.4	H_2O	1			
Ni	341.5	H_2O	3			

Desolvation	Remarks	Ref.
Yes	Compromise conditions	310
Yes		369
No		318

TABLE 7 (continued)

Element	Analysis line (nm)	Matrix	Detection limit (ng/ml)	HF generator Power (kW)	Frequency (MHz)	Nebulizer
Pb	405.8	H_2O	8			
Se	196.0	H2O	100			
Ti	334.9	H_2O	5	1	27	Pneumatic
V	437.9	H_2O	2			
Y	371.0	H_2O	0.6			
Zn	213.9	H_2O	10			
Al	396.1	H_2O	2			
Ce	456.2	H_2O	7			
Cr	357.8	H_2O	1			
Cu	324.7	H_2O	0.3			
Mn	403.0	H_2O	3			
Ni	351.5	H_2O	60			
Pb	405.7	H_2O	10			
W	400.8	H_2O	2			
Zr	349.6	H_2O	10			
Al	396.1	0.5% Fe solution	4	2.5	29	Pneumatic
Ce	456.2	0.5% Fe solution	13			
Cr	357.8	0.5% Fe solution	2			
Cu	324.7	0.5% Fe solution	0.3			
Mn	403.0	0.5% Fe solution	3			
Ni	351.5	0.5% Fe solution	70			
Pb	405.7	0.5% Fe solution	10			
W	400.8	0.5% Fe solution	3			
Zr	349.6	0.5% Fe solution	20			
Al	308.2	Lubri-cating oil	90			
Fe	259.9	Lubri-cating oil	40			
Pb	283.3	Lubri-cating oil	300	2.1	27	Pneumatic
Sn	284.0	Lubri-cating oil	30			
Zn	213.9	Lubri-cating oil	40			

Desolvation	Remarks	Ref.
No		318
Yes		368
No	Lubricating oil was diluted 1 : 10 w/v with methyl isobutyl ketone	347

prepared by additions of the analyte element to pure Fe solutions. Samples and blank reference solutions were measured alternately. No internal references lines were utilized. Calibration curves set up with the synthetic reference solutions were linear over a large concentration range, e.g. for Cr and Ni over more than four orders of magnitude. The limits of detection obtained for 12 elements in pure solution showed no significant deterioration in the presence of 0.5 wt.% Fe. They lay in the region between 0.06 and 0.001 $\mu g/ml$. The relative standard deviations for Al, Cr, Cu, Mn and Ni, in steel in a concentration range of % were found to be between 1.4 and 3.2%.

Scott and Kokot [317] described the application of the ICP to the analysis of geochemical samples. They used a crystal-controlled HF generator with a frequency of 27.12 MHz and a stabilized output power of 1 kW. The nebulizer was pneumatic without desolvation. The Jarrel Ash Monochromator Model No. 82000 in Ebert mounting had a focal length of 0.5 m and was equipped with a photomultiplier. The air dried soil samples were screened through an 80 mesh sieve. The 1 g samples were treated with 3 ml of a 9 : 1 mixture of concentrated $HClO_4$ and HNO_3 for one hour at 180°C. After cooling, the volume was made up to 50 ml. After removing the insoluble residue, the authors used the supernatant solution following decanting for their measurements without further dilution. For standards, they used solutions of the pure metal in HCl or HNO_3. For Cu, the calibration could be carried out with only an aqueous standard solution containing 37.5 $\mu g/ml$ and a blank solution, since the calibration curve was linear over four orders of magnitude. The conditions for Zn, Ni and Co were similar. Spectral interference was encountered only in the determination of Pb. The analytical values of the samples lay in the range 0.1—40 $\mu g/ml$ with the relative standard deviation 10% and below.

The determination of U in rocks (0.002—2% U_3O_8) with the same instruments has been described. The samples were crushed, pulverized and milled to 200 mesh. Solution of the samples took place in test tubes in a mixture of 5 ml 18 M HNO_3 and 5 ml of 1 + 1 H_2SO_4 which were heated until the H_2SO_4 started fuming. The volume of solution after cooling and filtering was made up to 100 ml. Aqueous standards were prepared from uranyl nitrate and used for calibration. For the spectral line at 378.28 nm, which has relative freedom from apparent spectral interference and a relatively low limit of detection (0.3 $\mu g/ml$), the optimal ICP parameters such

256

as RF power, coolant and support gas flow rates, and observation height in the plasma were used. Of the possible interfering elements, only Ca and Fe could occur in rocks in such a great concentration (>5%) as to cause a serious error. The Ca and Fe concentrations were therefore determined in the samples and the U values at below 0.02% were correspondingly corrected. A comparison of the values obtained with this direct analysis with mean values from four independent laboratories, each using a different technique, showed very good agreement. As a check, in further experiments the authors separated the U before determination with the aid of a 30 cm long ion exchange column (AG 1-X8, 100—200 mesh, Bio-rad Laboratories, Richmond, California). The exchange took place at pH 1—1.5. The successive eluates obtained with 100 ml of 0.1 M H_2SO_4 and 100 ml 6 M HCl were discarded. U was found in the last eluate with 80 ml 0.1 M HCl. This method did not yield more accurate results than the direct determination.

Scott and Strassheim [315] determined Fe, Mn, Cu, Al, B and Zn in the $\mu g/mg$ range in orchard leaves with the ICP-spectrometer system described above. The samples were dried for 24 h at 90°C and were burned to ash in silica crucibles at 500°C in a muffle furnace. The residue was evaporated with $1 + 1$ HNO_3 on a water bath and finally taken up with 4 ml $1 + 1$ HNO_3. The measurements on the solution, made up to 50 ml, were carried out against a pure aqueous standard solution. The agreement with the analytical results of other authors was satisfactory. It was only for Zn that an interference effect was observed.

Kniseley et al. [335] determined trace metals in μl volumes of biological fluids. 25 μl of blood, serum or plasma, has been used with essentially no sample preparation. A calibrated glass capillary served as a dosing system. A piece of rubber tubing with a pinch clamp and polyethylene capillary tubing linked the micropipette directly with the pneumatic nebulizer and connected desolvation unit. When the clamp was opened, the tube system filled with the blank solution, for which the nebulizer operated normally. When the clamp was closed, the flow stopped. An air bubble was sucked in by opening for a short time with the free end of the tube out of the solution. The micropipette containing the sample was then placed in the end of the tube and the other end of the glass tube was placed in the blank solution. On opening the clamp nebulization proceeded. The air bubble separated sample and blank solution so that the sample reached the plasma as a closed aerosol cloud. The limits of detection of the procedure lie for Mn at 0.002, for Cu at 0.006 and for Al at $0.007 \mu g/ml$.

In a similar microtechnique, Greenfield and Smith [68] dosed 1— $25 \mu l$ with micropipettes (Drummond "Microcaps") or microsyringes (Hamilton) by touching the end of the pipette to the inlet of the pneumatic nebulizer spraying into a small heated chamber. The ICP torch was connected directly to this chamber. The generator worked at 7 MHz giving a power of 0.5 kW into the plasma. The radiation was analyzed by either a Hilger and Watts Large Quartz Spectrograph or a FA 19 Polychromator. This method was used for the analysis of a number of matrices such as organic phosphorus compounds, blood and oil diluted with xylene. The calibration curves in the $\mu g/ml$ range are partly linear, partly slightly curved. The limits of detection for Cu, Cr, Fe, Ba, Al, La, Mg, Si, Ag, Pb and P lie in the range 10^{-9} to 10^{-10} g, the reproducibility at element quantities of 10^{-9} g is around 5%.

Fassel et al. [366] determined 15 different wear metals in lubricating oil with the ICP. The samples, reference and blank oils were diluted 1 : 10 w/v with MIBK and sprayed into the ICP. The calibration curves were linear over several orders of magnitude and no influence could be found of the viscosity on the measurement value. The limits of detection with an oil matrix lie between 0.4 ng/ ml and $0.3 \mu g/ml$, the relative standard deviations being a few percent, e.g. for $1.3—6 \mu g$ Mg/ml they lie between 2.5 and 7%.

Various organic compounds have also been determined with an ICP [355,370]. Since the discharge almost completely destroys organic compounds, it is, of course, not suitable for the production of the molecular spectra of larger molecules. On addition of hydrocarbons, the plasma shows lines and bands of elemental carbon and hydrogen, and of the fragments C_2 and CN, in the emitted spectra in the tail flame. The effect of different carrier gases such as air, O_2, N_2 or He was investigated, as were changes in the emitted spectra with the HF power and the spatial distribution in the plasma. (See also refs. 371—399.)

5. Stabilized d.c. arc plasmas and transferred plasmas

The plasma sources described in this section [5,17,34,61,62,105, 241,371,372,400—403] form a very heterogeneous group. They therefore have very different features of construction with many variations from the conventional d.c. arc to the free-burning flame-like plasmas. Because of this, their spectroscopic properties are very different. Since the main emphasis in this chapter on plasma excitation is on HF and UHF plasmas, not all the different kinds can be dealt with here in terms of construction and efficiency, but some of the important similarities and differences shall be established. Only some of the plasma sources serve as examples and the list is, of necessity, incomplete. Detailed descriptions of these plasma sources by many authors are to be found in the literature [5,17,34,372,404—432].

A feature common to all d.c. plasma sources is the d.c. arc discharge, which is stabilized in various ways or which is transferred away from the arc column, giving a flame-like appearance. The d.c. plasmas (DCP) may be split into two main groups [17]: (1) current carrying d.c. plasmas, among which may be differentiated (a) gas stabilized types and (b) disc or wall stabilized types; (2) current free d.c. plasmas, also called "plasmajets".

If a d.c. arc discharge is cooled in the peripheral regions by a gas stream, or is made narrow by walls, the electrical conductivity there goes down and the current is concentrated in the inner region of the plasma. Thus the arc channel is constricted and the increase in current density causes an increase in temperature. When the current density is high enough, this effect, the "thermal pinch" [5,34,38,241], causes a further magnetohydrodynamic pinch by the self-induced magnetic field. This constricts the discharge even more, and in this way the d.c. arc gains stability and a higher energy density. In particular, the wandering and flickering of the plasma column found with conventional arc excitation is avoided to a greater or lesser extent. DCPs reach very high temperatures in this way, e.g. Burhorn et al. [433] reached 50,000 K with currents up to 1500 A.

At temperatures above 5000 K, the plasma has a high viscosity and the effective injection of the sample into the plasma becomes very difficult [5,17,34,94,434—438]. Current carrying arcs set up an especially strong resistance to the introduction of aerosol particles because of the cooling effect of the impinging flow. With current free plasma jets, the mixing of plasma and aerosol is easier.

In what follows, various types of d.c. plasma will be introduced by some examples.

(A) INSTRUMENTATION

(1) Current carrying d.c. plasmas

Figure 18 gives, in schematic form, an example of a wall or disc stabilized d.c. plasma. Between anode and cathode are to be found several water-cooled metal discs, which are electrically isolated from each other and leave a narrow channel, in which the plasma burns, free between the electrodes. The number of discs, their separation and the diameter of the channel vary. Introduction of the aerosol is either through a hole in an electrode, or, as shown in Fig. 18, a special stabilizing disc. The emitted light emerges at the side between two discs or through a borehole in one of the electrodes.

The aerosol can be produced by well-known methods for all the d.c. plasmas. These are described in Sect. 3.B(1) and 4.B(2). Normally Ar is used as the working gas, but other noble gases, N_2, or air are also possibilities. The gas temperatures of the source depend on the power at which the arc is maintained, as well as on the construction of the instrument. At relatively low currents between 3 and 20 A and voltages between 90 and 220 V, they reach 6000–15,000 K [414,431]. Spectrochemical light sources of this kind have been described many times [401,414,416,420,427,431,434,439–445]. Zheenbaev et al. [439] describe a disc stabilized d.c. plasma jet in which the thoriated tungsten cathode is also formed as a disc with

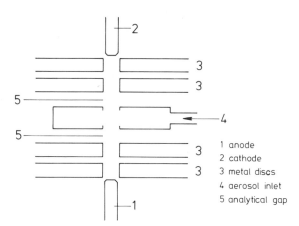

Fig. 18. Schematic diagram of a Marinković wall stabilized d.c. arc.

a central cavity. An argon stream carries the plasma upwards through this opening and an approximately 5 cm long plasma flame appears, which forms the actual light source.

The low temperature arc of Marinković et al. [417,419,426,446—448] is similar in construction to the one shown in Fig. 18. In addition to the stabilizing discs, a slow stream of Ar is introduced in the sections around the electrodes. The aerosol is introduced tangentially through a cavity in one of the stabilizing discs, which is not water cooled. The low temperature of about 3100 K was reached by the authors by adding KCl as a buffer to the analysis solution. In this way, the power of detection of this light source was increased by 1—3 orders of magnitude and lies in the lower ng/ml region. In the determination of impurities in high purity gold, the limits of detection in 0.25 M KCl solution lay between 0.001 and 0.6 μg/ml for 15 elements [448]. (See also the DCP of Owen [449], below).

(2) Current free d.c. plasma sources

In the current free plasma jets, the plasma is carried away from the arc column by a gas stream, thus forming a free-burning plasma flame in which no current is flowing.

In 1959, a light source was introduced for solution spectroscopy by Margoshes and Scribner [38] and independently by Korolev and Vainshtein [39] (see also refs. 40—42, 425, 450, 451). Its principle is shown in Fig. 19. Between the two ring-shaped carbon or graphite electrodes, a d.c. arc burns in a chamber with an arc current ranging from 15 to 25 A. The stabilizing gas, Ar or He, is introduced

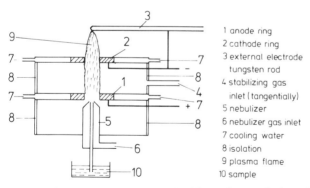

1 anode ring
2 cathode ring
3 external electrode
  tungsten rod
4 stabilizing gas
  inlet (tangentially)
5 nebulizer
6 nebulizer gas inlet
7 cooling water
8 isolation
9 plasma flame
10 sample

Fig. 19. Schematic diagram of the Margoshes et al. d.c. plasma jet.

tangentially into the space between the two electrodes, and the plasma is transferred through the ring cathode by the vortex coolant flow. The plasma flame, which therefore appears above the ring cathode, reaches a height of 1—2 cm and forms the actual radiation source.

The effective introduction of the aerosol into the plasma presents a great problem in this construction as well. The analysis solution is sprayed with a direct nebulizer through the annular anode into the space between the two electrodes. The atomizer, which is electrically isolated, is operated with argon and cooled with water, as are the ring electrodes.

An additional external cathodic electrode, a tungsten rod, 3, was introduced by Owen [449] to improve the stability. The arc discharge now burns between the ring anode and the rod cathode which is touched by the plasma flame. In this way, a current flows once again in the outer, visible part of the flame. The exit orifice electrode is not involved in the arc discharge.

Similar plasma sources and their properties have been repeatedly described in the literature [403,409—411,418,432,435,449—456], but all of them still have disadvantages [5,17,401]. Sources of interference are to be found, especially in the partly unsatisfactory stability of the arc, which wanders about on the electrode surfaces. The electrodes, even if they are only slightly eroded by the arc [449], contribute to the spectral emission as do coolant and carrier gas. The mixing of the aerosol into the plasma is not satisfactory, so that the optimum possible power of detection is usually not reached.

The plasma jet of Valente and Schrenck [409] consists of two separate chambers for the anode and cathode, which, after ignition of the arc, form an angle of 30° with each other (see Fig. 20). The chambers of aluminium tubing with a diameter of 10 mm have, at the upper end, an electrically isolated water-cooled control orifice, while the other end is sealed and carries the water-cooled electrodes. Under the control orifice, the sample aerosol with Ar (1.2 l/min) is blown into the anode chamber. A considerably slower Ar gas flow (10—50 ml/min) enters the chamber near the anode to prevent the incoming sample from contacting the hot alumina tube. The cathode chamber is a mirror image in its arrangement, with the difference that only Ar without sample aerosol is blown into it.

A d.c. arc is initiated by extending the graphite cathode rod into the anode chamber to contact the anode. After the arc is ignited, the

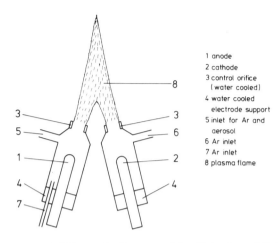

1 anode
2 cathode
3 control orifice
  (water cooled)
4 water cooled
  electrode support
5 inlet for Ar and
  aerosol
6 Ar inlet
7 Ar inlet
8 plasma flame

Fig. 20. Schematic diagram of the Valente and Schrenck d.c. plasma jet.

rod is withdrawn and the two chambers are turned to the position shown in Fig. 20, so that the argon carries the plasma out of the chambers, and the two plasma tongues unite to form one flame in which, however, the sample is unsymmetrically distributed. Since directly injected aerosol quenches the discharge and wet aerosol from an indirect nebulizer results in poor sensitivity, sample desolvation is necessary.

Yamamoto [406—408], with his plasma jet construction (shown in Fig. 21) [429,457—462], attempted to solve the problem of electrode erosion. The arc discharge burns inside the conical cavity of the water-cooled copper anode, 2, into which was placed the thoriated tungsten cathode, 1. Argon, which is used as discharge gas, is introduced tangentially, 5, and carries the plasma upwards and out through an opening where it appears as a flame. The sample is introduced obliquely from above through a direct nebulizer, 3, or through an aerosol conductor from an independent atomizer with a second Ar flow. This gas flow strikes the side of the arc column, pushing the anode foot of the arc against the opposite side of the anode nozzle to the position shown in Fig. 21. In this kind of gas introduction, the sample cannot reach the cathode or anode foot and cause corrosion there, but is quickly taken from the discharge zone into the flame.

Kranz   [422,423,437,438,463—465]   investigated   the   basic

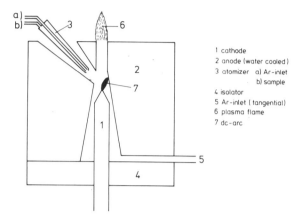

1 cathode
2 anode (water cooled)
3 atomizer a) Ar-inlet
           b) sample
4 isolator
5 Ar-inlet (tangential)
6 plasma flame
7 dc-arc

Fig. 21. Schematic diagram of the Yamamoto d.c. plasma jet.

methods of introducing substances into arc plasma and developed his current free, transferred type plasma jet from a wall stabilized arc (see Fig. 22). Between two horizontally arranged electrodes, 1, 2, burns a d.c. arc which is stabilized by water-cooled copper discs, 3, brought right in front of the electrodes. A gas stream, 5, (N_2, Ar, He or air are used), together with the sample aerosol, is blown into the middle of the arc vertically, from below. This carries the plasma as a long flame, 7, out through the orifice, 8. The thin laminar plasma flame reaches a length of 10—40 cm, depending on the gas used. In

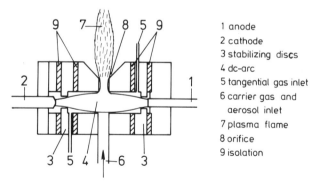

1 anode
2 cathode
3 stabilizing discs
4 dc-arc
5 tangential gas inlet
6 carrier gas and
  aerosol inlet
7 plasma flame
8 orifice
9 isolation

Fig. 22. Schematic diagram of the Kranz d.c. plasma jet.

264

order to prevent corrosion of the electrode material caused by reaction with the plasma flame, a protective gas is blown in tangentially through the stabilizing discs, 3, expelling the electrode vapours through radial outlets in the copper discs.

(3) Additional instrumentation

The mains instruments needed to run the various d.c. plasma sources are the same power supplies as used for conventional arc excitation. They will, therefore, not be described further here. The power of these instruments depends on the requirements of the particular excitation source and covers a wide range. Currents between about 5 A [419,426,448] and 500 A [429,460] and voltages between 15 V [406—408] and 250 V [466] are used, but usually the values lie around 15—25 A and 80—120 V. The gas consumption is usually between 4 and 20 l/min, with extreme values of 40 l/min [456].

For the production of the aerosol [434,436—438,456,462] all the types of nebulizers described in Sects. 3 and 4 may be used. Direct nebulization into the plasma [38,400,418,449,456, 460,467, 468], indirect nebulization with a spray chamber [419,426,431,432, 469—471], pneumatic and ultrasonic nebulization [455] with and without desolvation [409,414,435,442,455] are described in the literature.

Zheenbaev et al. [439] inject solutions directly into the plasmas in the form of a thin jet of liquid, which is pressed through a capillary 0.03—0.05 mm in diameter (see also ref. 472). It has also been tried with varying degrees of success to introduce powders and solids into the plasma [401,409,423,439,443,453,473—480].

(B) PROPERTIES AND INTERFERENCE

The temperatures of the plasma flames [414,425,435,440,441, 444,473,481] generally lie between 6000 and 12,000 K, but higher values have also been given. There is an exception in the low temperature arc with a KCl buffer, as used by Marinković et al. [419,426,448] at about 3100 K. The spectrum and the intensity of the spectral lines and of the spectral background are heavily dependent on the type of plasma burner used and on the operational parameters. There is usually a certain spatial distribution of the analyte in the plasma to be observed [482].

Generally, an increase in current increases the intensity of all lines.

The electrodes may form a source of error [429,460] since they erode to a greater or lesser extent in almost all types of jet and their material is then noticed in the spectrum. The electrodes must be changed several times during operation, especially at high currents.

Matrix effects of various kinds and intensity have been observed by many authors [400,409,419,450,452,468,483]. In current carrying d.c. plasmas, easily ionizable elements influence the line intensities of many elements, a fact which can be made use of in buffering [409,419,426,450,456,484] and improving the limits of detection. These plasmas are not influenced by the introduced aerosol. In current free types, the conditions are exactly opposite [483].

Because of the high plasma temperatures, interference due to incomplete dissociation or the formation of thermally stable compounds is weak or nonexistent [429,460].

It is not possible to summarize the advantages and disadvantages of the DCPs because of the multitude of different constructions. According to Boumans [17] the results obtained with gas stabilized arcs are not satisfactory, since the problem of the optimal, precise introduction of the sample aerosol into the plasma is not completely solved [435] and the matrix effects caused by alkalis interfere.

The precision which can be reached with the individual DCP types are also very varied. The best values lie around 1% for the relative standard deviation [38,402,432], although considerably higher values of over 20% [452] have also been reported. The medium term precision within periods of hours is bad [17,241].

The limits of detection of the d.c. plasmas lie, in general, in the $\mu g/$ ml and ng/ml range. Instruments with Echelle spectrometers with their high resolving power, show considerably increased powers of detection [17]. A list of such data for various instruments has been drawn up in Table 8. However, a real comparison of the efficiencies of the individual types cannot be made from this table since the instrumental conditions under which these data were gained are very different.

The performance data are certainly not sufficient at present to allow the DCPs to compete with HF and UHF plasmas such as ICPs and CMPs, but there is a very real possibility of further developments and improvements in some of these spectroscopic light sources, so

266

TABLE 8

Some detection limits for different types of d.c. plasma sources for aqueous solutions

Element	Analysis line (nm)	Detection limit (μg/ml)	Type of source	Ref.
Al	396.2	0.21		
B	249.7	0.013		
Co	345.3	0.23		
Cr	357.9	0.042	Marinković	439
Fe	372.0	0.11		
Mg	285.2	0.035		
Ni	352.4	0.23		
Zn	213.9	0.018		
Al	396.2	0.3		
As	235.0	0.5		
B	249.8	0.05		
Bi	306.8	5		
Co	228.6	0.8		
Cr	267.7	0.4		
Cu	324.8	0.2		
Fe	259.9	0.2		
Hg	254.6	0.3		
Mg	279.7	0.02		
Mn	257.6	0.04		
Mo	281.6	0.1	Owen	432
Ni	221.6	0.3		
Pb	283.3	0.2		
Se	204.0	2		
Si	251.6	0.2		
Sn	284.0	0.6		
Te	214.3	0.7		
V	309.3	0.2		
W	400.9	2		
Zn	213.9	0.1		
Zr	339.2	0.3		
B	249.8	0.4		
Cd	228.8	0.03		
Cr	425.4	0.003		
Fe	372.0	0.005	Valente—	409
Ni	341.5	0.003	Schrenck	
Pb	283.3	0.03		
U	424.2	0.5		
Zn	213.9	0.01		

TABLE 8 (continued)

Element	Analysis line (nm)	Detection limit (μg/ml)	Type of source	Ref.
Ag	405.7	1.3		
Al	396.2	2		
B	249.7	0.2		
Cd	228.8	0.7	Yamamoto	429
Co	228.0	5		
Cr	357.8	3		
Cu	324.7	1.5		
Al	396.2	0.2		
Ca	422.7	0.1		
Cu	510.6	1.0		
Fe	438.3	2.0	Kranz	464
Mg	518.4	0.2		
Mn	403.1	0.2		
Zn	481.1	20.0		

that they also remain of interest for future analytical purposes, especially as the investment and running costs are relatively low.

(C) PRACTICAL APPLICATION

Doerffel and Lichtner [431] described the use of a disc stabilized d.c. arc for spectrometric solution analysis. They determined zinc in copper alloys and silicon in oxide samples in the % range.

Zmbova and Marinković [448] used a gas stabilized low temperature arc (see Fig. 18) with aerosol supply for the determination of 14 elements present as impurities in high purity gold. They compared two analytical procedures, the direct aspiration of the dissolved sample and the determination of the separated elements after the ether extraction of gold from 1 M HCl. In both cases, KCl (0.25 M) was added to the solutions as a spectroscopic buffer. For Fe, Al and Cr, the relative standard deviation obtained was between 1.5 and 6% for 0.005—0.03 μg/ml.

268

Goto and Atsuya [460] determined Al and B in steel with a d.c. jet as developed by Yamamoto (see Fig. 21). The Fe was removed by extraction with MIBK after the sample was dissolved. In a multistep separation procedure, Al and Be were isolated from other accompanying elements and determined by a solution spectrometric procedure. For Al contents of 0.06%, the relative standard deviation lies at 4.9%. The authors also determined Ca and Mg in iron ores and cast iron [458] and the rare earths [459].

The analysis of geological samples with various d.c. plasma jets has been described several times [400,450,451,485]. Golightly and Harris [400] determined Al, Ca, Fe, Si and Ti in the % range in geochemical reference samples, and Ti in lunar basalts, gabbros and soils. The general accuracy of analysis was within approximately 10% of the accepted values, and relative standard deviation ranged from 2 to 6%. The authors give an extensive bibliography.

Corcoran et al. [485] used a commercial d.c. jet, the Spectra-Span Model 101 from Metrametrics, Inc., for the rapid determination of Ca and P in phosphate rock. After fusing the sample with anhydrous $LiBO_2$ for 10 min and dissolving the melt in 20% HNO_3, the solution was analyzed without any separation. The same light source was also used for the investigation of the composition of coal ash [476].

The concentration of F, S, P, B, Zn and B in rocks was determined by Savinova and Karyakin [450,451] in the range 10^{-4}—1%. The well ground samples were pressed into carbon electrodes which were used as anodes in a d.c. plasma jet. The relative standard deviation was 15% for the average concentration being determined. The limits of detection for this solid analysis were: B 2.5×10^{-4}%; S 5×10^{-2}%, P 2.5×10^{-3}%, Zn 2.5×10^{-3}% and Be 5×10^{-5}%.

For further literature in connection with the use of this source, see refs. 486—492.

6. Other HF plasmas

In addition to the HF and UHF plasmas dealt with in Sects. 3 and 4, a range of other HF plasma sources is described in the literature, which in their construction or method of energy coupling depart from the construction principles of the ICP and CMP to a greater or lesser extent, and which are used for various purposes. Their importance in solution spectral analysis is not as great as that of ICP or

CMP. Since there are also many different kinds, a short and incomplete list must suffice here so as to remain within the bounds of this chapter. Some inductively coupled HF plasmas have also been included in this list, although in the literature they are also described as ICP sources. This nomenclature is, of course, correct. Its important difference from the "real" ICP in the form developed by Greenfield et al. [61,62,66] is that the flows of coolant gas and carrier gas are missing. The plasma gas is, therefore, fed through a single glass or silica tube, in which the plasma burns inside the coil, spread out to a greater or lesser extent. In the framework of this survey of plasma excitation, however, the name ICP shall be used only for the burner type developed by Greenfield, since it is only with this that high efficiencies in solution spectrometry are reached.

As well as inductive coupling, capacitive coupling is also used to produce HF plasmas. For this, an electrode is needed for the power transmission, as described for the CMP. In a wide-ranging state of the art review with 131 references, Eckert [270] describes the historical development, techniques, theory and applications of HF induced plasmas.

A high frequency low wattage discharge (40 MHz, 200 W) with Ar at atmospheric pressure as radiation source with inductive coupling is described by Kleinmann et al. [337,493—495]. In the discharge chamber, two water-cooled tantalum holders grip a graphite support and conduct the heating current. The holders are inside a silica tube, the HF coil being arranged concentrically round the outside. The argon passes through the silica tube. Sample volumes of 30 μl are dried and evaporated from the graphite support by ohmic heating. The plasma burns above the graphite disc at atmospheric pressure, but can also be maintained at reduced pressure. Spectral and physical properties of the plasma and detection limits, which are in the range 0.1—10 ng for 30 μl samples [337], have been investigated [493—495].

Johnston and Lawton [496] studied the properties of a premixed propane—air flame which was augmented by inductive coupling of HF power.

Runser and Frank [497,498] studied the emission spectra of organic molecules such as chlorinated pesticides in a 13.56 MHz discharge. Argon was used as a carrier gas. The plasma cell, made of pyrex glass, was positioned axially through the tuning coil. The tube, which was sealed at one end with a silica window, was divided by means of a small nozzle into a high pressure section and a low pressure section. The sample inlet, a tube bent into a right-angle and in front of which was placed a tantalum strip, was in the high pressure section. The low pressure section was connected to a vacuum pump. A characteristic plasma flame and striated discharge for use as an emission source for spectrochemical analysis were contained in the low and high pressure regions of the cell, respectively.

The source described by Dunken et al. [49,52—54,344,499,500] as "HF torch discharge" is a capacitively coupled HF plasma. The discharge, initiated by a 50 MHz generator burns above a water cooled electrode formed as a copper tip at the end of a Lecher conductor. This electrode is located in a glass tube distended to a sphere, through which pressurized air or nitrogen and the sample aerosol flow. In another construction [54], one end of a HF coil is earthed while the other, the so-called "hot end", is in the form of a water cooled metal tip and carries the discharge. Similar and other capacitively coupled HF plasmas with their properties and possible applications have been described by various authors [46,47,50,51,81, 501—506].

A combined inductive and capacitive coupling of the HF power with the plasma is used with some excitation sources [55,82,103, 251,501]. Again, the free end of a HF coil forms the electrode tip, on which the plasma sits. The electrode is found inside a silica tube, round which the HF coil is arranged concentrically (see Fig. 23). The sample aerosol, produced by a pneumatic or ultrasonic nebulizer, is either fed inside the electrode [55] or flows along the silica tube on the outside of the electrode. N_2 [103], He [55,257], or Ar serve as plasma gas. The limits of detection obtained with this excitation source lie in the ng/ml range for pure aqueous solutions [103].

Cristescu and Giurgea [48,101,102] superimposed a d.c. discharge on a capacitively coupled HF discharge by introducing an additional electrode from a d.c. voltage source at various heights in the plasma torch. The second d.c. pole was connected by a choking coil with one of the coils in the HF oscillating circuit. The HF discharge burned between two condenser plates in the oscillating circuit, of which one was in the form of a small disc with a metal tip mounted on it as an electrode for the discharge, and the other was plate or ring shaped.

Another HF plasma source (27.5 MHz, 250 W) for emission spectroscopy in the vacuum ultraviolet range is described by Dreher

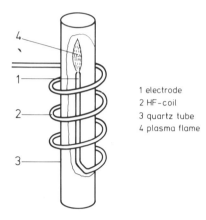

1 electrode
2 HF-coil
3 quartz tube
4 plasma flame

Fig. 23. Schematic diagram of the d.c. plasma jet of an inductively and capacitively coupled plasma torch.

and Frank [507]. The basic discharge chamber consisted of a pyrex glass tube with a grounded tantalum foil inside. The plasma cell was inside the HF coil and was connected, vacuum tight, with the spectrograph. The sample solution was dosed onto the tip of the tantalum strip and dried in a stream of warm air. The discharge tube was filled with Ar and after the flow was stopped, the UHF discharge was initiated. One end of the moving striated plasma was attracted to the grounded tantalum foil, heating the tip to vaporize the sample. Wilson et al. [508,509] and Botschkova et al. [510] used high frequency discharges in gas capillaries with external electrodes as light sources in the spectral analysis of organic and inorganic gases and gas mixtures. This excitation source, also known as Tesla discharge, is maintained at reduced pressure. The spectra contain strong atomic lines as well as the molecule bands.

A further electrodeless HF discharge in a water-cooled discharge tube, similar to a Liebig condenser, with external ring electrodes was used by Egorova [249] for solution analysis. The aerosols, produced by pneumatic or ultrasonic nebulizers, were transported into the discharge zone with Ar (2 l/min). Two forms of discharge occurred depending on the ionization potential and upon the concentration of the elements in the aerosol. One form of discharge showed an individual structure, while the other filled the whole discharge chamber.

(See also refs. 511—519.)

272

7. AAS with plasma sources

For some elements, combustion flames are not suitable for producing the atomic gas that is necessary for AAS or AFS in order to produce a sufficiently strong analysis signal. There are chiefly three reasons for this [271].

(a) Because of the relatively low temperatures and short residence times, elements which are not easily atomized and thermally stable compounds are not atomized sufficiently so that the number of atoms in the vapour state is too low.

(b) Due to chemical reactions in the flame, compounds form which are difficult to dissociate thermally.

(c) The nebulization and the sample introduction into the flame are not very efficient.

Since most of the plasma excitation sources (ICP, CMP and d.c. plasmas) have a higher temperature than combustion flames, can also be run with inert gases such as Ar or N_2, and the residence time of the atoms in the plasma is longer than that for a flame, it was logical to try to use these plasmas as atom cells in AAS and AFS [271,457, 471,520—531] in order to provide a higher analyte-free atom concentration in the vapour phase. In particular, ICPs possess a further advantage over the flame because of their higher efficiency of aspiration, volatilization and atomization [528]. The higher gas temperature of the plasma does, however, mean that the ionization equilibrium also moves to the side of the ions, so that the number of free atoms again falls. By addition of easily ionizable elements such as the alkali metals and alkaline earths, the ionization may be considerably suppressed here as well.

An important factor, which greatly reduces the use of the various plasmas as atom cells in AAS and AFS, is the complexity of the instrumentation, and therefore the financial disadvantage, which means that it is certainly only worth using them where chemical flames are of no use and give bad results, such as, for example, with some strong monoxide-forming elements, e.g. Al, Nb, Ti, W and Y [530]. In most cases, the flame will go on being used, while emission spectroscopy will be carried out by the plasma, to make use of the great advantage of simultaneous multi-element analysis which is not possible with AAS.

Views on the value of plasma for AAS and AFS are divided. Veillon and Margoshes [271] in 1968 were already of the opinion that "except for a few elements, ICP is not a suitable replacement for the chemical flames", particularly as the limits of detection are poor and the absorption of a high emission signal is superimposed. Also, Greenfield et al. [36] see little if any advantage in the use of an ICP in AAS. At first, Fassel et al. [528,530] were opposed to this view and maintained that these plasmas represented a valuable complement to the flames as atom reservoirs for AAS and especially in AFS. The lower spectral background and the fact that some matrix effects were not present, e.g. the effect of P and Al on the adsorption signal, were advantageous. However, in 1976, even Fassel saw no future for the ICP as an adsorption cell, especially since the emission is very intense [532].

In addition to the ICP burner with three concentric tubes described in Sect. 3 [271,528—530,533], various special types have been built for use in AAS and AFS. Greenfield et al. [65] described a T-shaped torch as atomizer, the stem of which consists of a plasma torch substantially of the usual form. The horizontal tube, the actual atom reservoir, can either be open at both ends or closed at one end by an optical flat.

A HF induced plasma with a single tube as an atomizer in AAS is also described by Bordonali and Biancifiori [534—539], the absorption chamber of which consists of a branch tube closed by silica windows. Morrison and Talmi [524,540,541] used a HF furnace for AAS and emission. Microsamples of solids or evaporated solutions were vaporized and atomized in a HF induction heated graphite crucible, over which a helium plasma was formed due to the same HF field, which was passed by the light of the hollow cathode lamp. The sample was introduced from above into the graphite crucible through a PTFE sample chamber which was flushed continuously with helium. The limits of detection obtained with this instrument lay between 10^{-8} and 10^{-11} g for 6 elements.

Bachurina et al. [542] determined small quantities of Ba, Cr and Ca in deposits in electronic tubes by an UHF plasma discharge with both emission and atomic absorption techniques. The limits of detection that they obtained lay in the ng range.

Marinković and Vickers [471] used a U-shaped long-path stabilized d.c. arc device to investigate the free atom production by AAS. The aerosol introduction was vertical tangentially to the arc. Measurements of the absorption signals were made up to 6 mm off the arc axis. The power of detection lay in the μg/g range.

274

Malinek and Massman [527] used a new design of the Kranz plasma jet with a multipass optical attachment as a source of atomic vapour in AAS. The authors indicate that this plasma appears especially suitable for elements like B or Se with resonance lines below 250 nm.

Various parameters are decisive for optimal measurement of the absorption. They include gas flow rate, power and observation height [528] which, especially in AFS, influence the analysis signal, the background [271,530] and therefore the signal-to-noise ratio. Since the gas temperature partly determines the state of ionization equilibrium, and since as many of the analyte atoms as possible should be in their ground state, exact temperature control is an advantage [65].

8. Comparison of the various plasma types

A comparison of the various plasma types described in this chapter by evaluating the published work is a daring venture, as is generally the case with any comparison of the efficiencies of different determination procedures or principles in analytical chemistry, and particularly in trace analysis, because it is very seldom that the published statistical data or information about interference have been obtained under exactly the same instrumental conditions for them to be compared objectively [94]. In spite of the multitude of existing data, such an investigation is necessarily subjective and depends on the author's personal preconceptions. His judgement reflects the requirement of his own field of work for an optimal procedure, and the individual facts are correspondingly weighed and valued. Add to this the fact that the plasma torches of any one kind, such as the ICPs but especially the DCPs, are different in construction and properties to a greater or lesser extent, and it is clear that a unified picture cannot be built up. The comparisons often found in the literature between spectroscopic plasma excitation procedures and other methods of analysis, such as AAS, also cannot always be objective, and often, if a different set of data are used, a contradictory conclusion may result.

Since it is often the case that in the description of new procedures efficiencies are presented in a very optimistic light, a certain pessimism can also be quite useful when evaluating the literature.

TABLE 9

Comparison of the different plasma excitation sources

	ICP	CMP	MIP	DCP
Instrumental complexity	+ −	+	+ +	+
Running costs	−	+	+ +	+
Efficiency; overall evaluation	+ +	+	+ −	+ −
Powers of detection	+ +	+	+ +	+ −
Total susceptibility to interference	+ +	+	−	−
Matrix effects	+ +	+ −	−	+ −
Stable compound formation	+ +	+	−	+ −
Ionization interference	+ +	−	−	−
electrode material	+ +	+ −	+	−
Efficiency of sample introduction	+ +	+ −	+	−
Possibility of simultaneous multi-element analysis	+ +	+ +	+ −	+ +
Operation	+ −	+	+	+ −

Evaluation: Compared with the other excitation sources, the source under consideration is ++, very good; +, good; +−, neither good nor bad; −, bad.

Taking into account these problems, an effort has been made to compare the four most important plasma excitation sources. The comparison is presented in Table 9 with a four-point scale of evaluation.

In the comparison of the cost of investment and running of the instrument, the MIP certainly comes out best with its low power generator and very low gas consumption. Even for a CMP, the investment costs are lower than for an ICP, in which, of course, the generator power is an important factor. For the running costs, the very high gas consumption of the ICP needs to be considered. This is considerably lower in the CMP or DCP.

Under the heading of "efficiency; overall evaluation" are banded the topics power of detection, interference, etc. It is clear that the ICP has first place here, followed by CMP. In spite of its very good power of detection, at the present state the MIP cannot be judged generally as a good excitation source because of its susceptibility to interference. This will possibly be changed by further investigation, using the cavity of Beenakker [116].

The power of detection of the MIP reaches that of FAAS for many elements. ICPs have a better detection power for many elements than CMPs, especially when easily ionizable elements are absent. At the bottom of the list come the DCPs. These statements are, however, not valid for all elements, and for some the list is different.

The ICP is, without doubt, the excitation source which is least susceptible to interference. This goes both for influencing of the signal by matrix and accompanying elements and for changes in analysis line intensities due to easily ionizable elements. Today, the MIP is most susceptible to these sources of interference. In the CMP and the DCP there is additional interference due to the electrode material.

The efficiency of sample introduction is highest in ICPs, lowest in DCPs. In CMP and DCP, the sample aerosols may easily evade the plasma resulting in a low power of detection and an inconstant analysis signal. Simultaneous multi-element analysis is possible with ICP, CMP and DCP, but with MIP this possibility exists only to a certain extent with some limitations.

The heading "operation" includes everything in connection with the operation of the instruments during analysis, such as plasma ignition, optimizing and keeping constant the parameters, tuning, sample introduction etc. Here the MIP and CMP are certainly more easy to operate than the ICP or DCP.

In order to make a more pronounced evaluation, the analysis problem itself must be taken into account. For the reason above, it is not possible to declare one excitation source to be the best and decry the others' right to exist.

In the multitude of the problems of routine analysis, where many elements in a large number of similar samples of complex composition are to be determined in the shortest possible time as efficiently and as economically as possible, the ICP will certainly offer the optimal excitation source for emission spectroscopy. For special problems, however, like the determination of easily separable volatile elements such as Hg or compounds such as the metal hydrides of certain elements, the MIP may be far ahead of the ICP.

The CMP also finds meaningful areas of use, for example in extreme trace analysis where even the powers of detection of the ICP are not sufficient for a direct analysis and the elements must be separated and concentrated before determination, or if strongly defined matrices are present as in the blank value control of reagents. Also, the DCPs, which according to the table come off worst, are suitable for many analytical problems.

Various authors have carried out many comparisons of plasma excitation sources with each other or with other excitation and determination procedures [12,17,18,34—36,61,62,74,94,241,273,274, 303,310,367,384,543—545]. A real comparison of ICP and CMP was undertaken by Boumans et al. [94], where all the experimental conditions, with the exception of the excitation sources, were identical, involving standardized experiments and standard measuring equipment. The authors determined powers of detection, sensitivity and precision for 14 elements and investigated matrix effects produced by Cs_2SO_4, $CdSO_4$ and $(NH_4)_2HPO_4$. The authors come to the conclusion that the ICP is better than the CMP, which was indicated by (1) better power of detection, (2) smaller matrix effects, especially by ionization interferences, (3) higher sensitivity, (4) better reproducibility, and (5) the possibility of finding so-called "compromise conditions" for simultaneous multi-element analysis, which are suitable for many matrices and analytes.

In a similar investigation, Larson and Fassel [367] showed that the CMP is more susceptible to interference than the ICP.

The emphasis in the future development and applications of the plasma excitation source for emission spectral analysis of solutions will clearly lie on the inductively coupled high frequency plasma. But it would certainly be wrong if further research and development of the other plasma light sources were neglected.

References

1 G. Tölg, in G. Svehla (Ed.), Wilson and Wilson's Comprehensive Analytical Chemistry, Vol. III, Elsevier, Amsterdam, 1975.
2 G. Tölg, Talanta, 21 (1974) 327.
3 G. Tölg, Talanta, 19 (1972) 1489.
4 R.M. Barnes, in Dowden (Ed.), Emission Spectroscopy, Hutchinson and Ross, Strondsburg, Pa., U.S.A., 1976, p. 387.
5 P.W.J.M. Boumans, in E.L. Grove (Ed.), Analytical Emission Spectroscopy, Part 2, 6, Dekker, New York, 1972, p. 1.
6 P.W.J.M. Boumans, Theory of Spectrochemical Excitation, Hilger, London and Plenum Press, New York, 1966.
7 L.R.P. Butler, H.G.C. Human and R.H. Scott, in J.W. Robinson (Ed.), Handbook of Spectroscopy, Vol. 1, CRC Press, Cleveland, 1974, p. 816.

8 P.M. Chung, L. Talbot and K.J. Touryan, Electric Probes in Stationary and Flowing Plasmas: Theory and Application, Springer, New York, 1975.

9 R. Mavrodineanu and R.C. Hughes, Developments in Applied Spectroscopy, Vol. 3, Plenum Press, 1964, p. 305.

10 P.W.J.M. Boumans, Proc. Anal. Div. Chem. Soc., 14 (1977) 143.

11 P.W.J.M. Boumans and F.J. de Boer, XVI. Coll. Spectrosc. Int., Vol. II, 1971, p. 431.

12 P.W.J.M. Boumans and F.J. de Boer, Proc. Anal. Div. Chem. Soc., 12 (1975) 140.

13 P.W.J.M. Boumans, 20th Coll. Spectrosc. Int. and 7th Int. Conf. Atomic Spectrosc., Prague, September 1977.

14 P.W.J.M. Boumans, Eur. Spectrosc. News, 1 (1975) 13.

15 I. Kleinmann and B. Polej, Chem. Listy, 65 (1971) 1.

16 J.D. Winefordner, J.J. Fitzgerald and N. Omenetto, Appl. Spectrosc., 29 (1975) 369.

17 P.W.J.M. Boumans, Z. Anal. Chem., 279 (1976) 1.

18 P.W.J.M. Boumans, Philips Tech. Rundsch., 34 (1974/75) 306.

19 S. Greenfield, Proc. Anal. Div. Chem. Soc., 13 (1976) 279.

20 H. Meinke and F.W. Gundlach, Taschenbuch der Hochfrequenztechnik, Springer, Berlin, 1962.

21 E. Meyer and R. Pottel, Physikalischen Grundlagen der Hochfrequenztechnik, Verlag Friedr. Vieweg und Sohn, Braunschweig, BRD, 1969.

22 R.E. Collin, Grundlagen der Mikrowellentechnik, VEB Verlag Technik, Berlin, 1973.

23 A.J. Baden Fuller, Microwaves, Pergamon Press, Oxford, 1969.

24 H. Püschner, Wärme durch Mikrowellen, Philips Technische Bibliothek, 1964.

25 H.H. Klinger, Mikrowellen, Grundlagen und Anwendungen der Höchstfrequenztechnik, Verlag Radio-Foto-Kinotechnik, Berlin, 1966.

26 W.B. Kunkel, Plasma Physics in Theory and Application, McGraw-Hill, New York, 1966.

27 S.C. Brown, Basic Data of Plasma Physics, MIT Press, Cambridge, Mass., 1956.

28 M.A. Heald and C.B. Wharton, Plasma Diagnostics with Microwaves, Wiley, New York, 1965.

29 J.R. Hollahan, in A.T. Bell (Ed.), Techniques and Applications of Plasma Chemistry, Wiley, New York, 1974.

30 B.T. Tolok (Ed.), Recent Advances in Plasma Diagnostics, Consultant Bureau, New York, 1971.

31 F.K. McTaggert, Plasmachemistry in Electrical Discharges, Elsevier, Amsterdam, 1967.

32 J.D. Swift and M.J.R. Schwar, Electrical Probes for Plasma Diagnostics, Elsevier, Amsterdam, 1970.

33 A.P. Thorne, Spectrophysics, Halsted, New York, 1974.

34 S. Greenfield, H.McD. McGeachin and P.B. Smith, Talanta, 22 (1975) 1.

35 S. Greenfield, H.McD. McGeachin and P.B. Smith, Talanta, 22 (1975) 553.

36 S. Greenfield, H.McD. McGeachin and P.B. Smith, Talanta, 23 (1976) 1.

37 H. Gerdien and A. Lotz, Wiss. Veroeff. Siemens-Konzern, 2 (1922) 489.

279

38 M. Margoshes and B.F. Scribner, Spectrochim. Acta, 15 (1959) 138.
39 V.V. Korolev and E.E. Vainshtein, J. Anal. Chem. (USSR), 14 (1959) 658.
40 V.V. Korolev and E.E. Vainshtein, J. Anal. Chem. (USSR), 14 (1959) 731.
41 V.V. Korolev and E.E. Vainshtein, J. Anal. Chem. (USSR), 15 (1960) 686.
42 V.V. Korolev and E.E. Vainshtein, J. Anal. Chem. (USSR), 15 (1960) 783.
43 G. Cristescu and R. Grigorovici, Bull. Soc. Roum. Phys., 42 (1941) 37.
44 G. Cristescu and R. Grigorovici, Opt. Spectrosk., 6 (1959) 129.
45 G. Cristescu and R. Grigorovici, Naturwissenschaften, 29 (1941) 571.
46 A.L. Stolov, Uch. Zap. Kazan. Gos. Univ., 116 (1956) 118.
47 E. Badaran, M. Giurgea, Ch. Giurgea and A.T.H. Trutia, Spectrochim. Acta,
    11 (1957) 441.
48 G.D. Cristescu, Ann. Phys., 6 (1960) 153.
49 H. Dunken, W. Mikkeleit and W. Kniesche, Acta Chim. Acad. Sci. Hung.,
    33 (1962) 67.
50 W. Tappe and J. van Calker, Z. Anal. Chem., 198 (1963) 13.
51 V. Trunecek, Z. Chem., 4 (1964) 358.
52 H. Dunken, G. Pforr and W. Mikkeleit, Z. Chem., 4 (1964) 237.
53 G. Pforr and K. Langner, Z. Chem., 5 (1965) 115.
54 H. Dunken and G. Pforr, Z. Phys. Chem., 230 (1965) 48.
55 R. Mavrodineanu and R.C. Hughes, Spectrochim. Acta, 19 (1963) 1309.
56 G.J. Babat, Vestn. Electroprom., (1942) No. 2, 1; No. 3, 2.
57 G.J. Babat, J. Inst. Electr. Eng., 94 (1947) 27.
58 T.B. Reed, J. Appl. Phys., 32 (1961) 821.
59 T.B. Reed, J. Appl. Phys., 32 (1961) 2534.
60 T.B. Reed, Int. Sci. Technol., June (1962) 42.
61 S. Greenfield, J.L. Jones, C.T. Berry, R.H. Wendt and V.A. Fassel, Analyst,
    89 (1964) 713.
62 S. Greenfield, J.L. Jones, C.T. Berry, R.H. Wendt and V.A. Fassel, Brit.
    Pat. 1,109,602, 1968.
63 S. Greenfield, ICP Inf. Newsl., 1 (1) (1975) 3.
64 S. Greenfield, J.Ll.W. Jones, C.T. Berry and L.G. Bunch, Proc. Soc. Anal.
    Chem., 2 (1965) 111.
65 S. Greenfield, P.B. Smith, A.E. Breeze and N.M.D. Chilton, Anal. Chim.
    Acta, 41 (1968) 385.
66 S. Greenfield, I.Ll.W. Jones and C.T. Berry, US Pat. 3,467,16, 1969.
67 S. Greenfield and P.B. Smith, Anal. Chim. Acta, 57 (1971) 209.
68 S. Greenfield and P.B. Smith, Anal. Chim. Acta, 59 (1972) 341.
69 S. Greenfield, P.B. Smith and H.McD. McGeachin, Proc. Soc. Anal. Chem.,
    10 (1973) 89.
70 S. Greenfield, Eur. Spectrosc. News, 1 (1976) 4.
71 R.H. Wendt and V.A. Fassel, Anal. Chem., 37 (1965) 920.
72 V.A. Fassel and G.W. Dickinson, Anal. Chem., 40 (1968) 247.
73 W. Barnett and V.A. Fassel, Spectrochim. Acta Part B, 23 (1968) 643.
74 G.W. Dickinson and V.A. Fassel, Anal. Chem., 41 (1969) 1021.
75 H.P. Broida and J.W. Moyer, J. Opt. Soc. Am., 42 (1952) 37.
76 H.P. Broida and M.W. Chapman, Anal. Chem., 30 (1958) 2049.
77 M. Yamamoto and S. Murayama, Spectrochim. Acta Part A, 23 (1967) 773.

78 M. Yamamoto, Ger. Pat. 1,208,100.
79 J.W. Runnels and J.H. Gibson, Anal. Chem., 39 (1967) 1398.
80 A.J. McCormack, S.C. Tong and W.D. Cooke, Anal. Chem., 37 (1965) 1470.
81 J.D. Cobine and D.A. Wilbur, J. Appl. Phys., 22 (1951) 835.
82 R. Mavrodineanu and R.C. Hughes, Dev. Appl. Spectrosc., 3 (1964) 305.
83 U. Jecht and W. Kessler, Z. Anal. Chem., 198 (1963) 27.
84 U. Jecht and W. Kessler, Z. Phys., 178 (1964) 133.
85 A. van Sandwijk and J. Agterdenbos, Kurzreferate Euroanalysis I, 1972, p. 98.
86 A. van Sandwijk, P.F.E. van Montfort and J. Agterdenbos, Talanta, 20 (1973) 495.
87 A. van Sandwijk and J. Agterdenbos, Talanta, 21 (1974) 360.
88 R.F. Browner and J.D. Winefordner, Spectrochim. Acta Part B, 28 (1973) 263.
89 Safety Level of Electromagnetic Radiation with Respect to Personnel, United States of America Standards Institute USA SI, 1966, C 95.1.
90 J.L. Stanley, H.W. Bentley and M.B. Denton, Appl. Spectrosc., 27 (1973) 265.
91 D.W. Golightly, ICP Inf. Newsl., 3 (1) (1977) 14.
92 P. Czerski, Commun. Int., 3 (1976) 32.
93 J.J. DeCorpo, J.T. Larkins, M.V. McDowell and J.R. Wyatt, Appl. Spectrosc., 29 (1975) 85.
94 P.W.J.M. Boumans, F.J. de Boer, F.J. Dahmen, H. Hoelzel and A. Meier, Spectrochim. Acta Part B, 30 (1975) 449.
95 K. Laqua, W.D. Hagenah and H. Wächter, Z. Anal. Chem., 225 (1967) 142.
96 P.N. Keliker and C.C. Wohlers, Anal. Chem., 48 (1976) 333A.
97 A. Danielsson and P. Lindblom, Appl. Spectrosc., 30 (1976) 151.
98 G. Kaiser, D. Götz, P. Schoch and G. Tölg, Talanta, 22 (1975) 889.
99 T. Kántor, in G. Svehla (Ed.), Wilson and Wilson's Comprehensive Analytical Chemistry, Vol. V, Elsevier, Amsterdam, 1975.
100 R. Mannkopf and G. Friede, Grundlagen und Methoden der chemischen Emissionsspektralanalyse, Verlag Chemie, Weinheim, 1975.
101 G.D. Cristescu and M. Giurgea, Z. Chem., 7 (1967) 360.
102 G.D. Cristescu and M. Giurgea, Opt. Spectrosk., 11 (1961) 424.
103 C.D. West and D.N. Hume, Anal. Chem., 36 (1964) 412.
104 R.M. Barnes, Anal. Chem., 44 (1972) 122R.
105 B.L. Sharp, Selected Annual Reviews of Analytical Science, Vol. 4, 1976, p. 37.
106 J. Kleinmann, Radioisotopy, 11 (1970) 295.
107 H. Hinkel, Magnetrons, Philips Techn. Bibliothek, 1961.
108 K.L. Smith, Rev. Sci. Instrum., 44 (1973) 1108.
109 F.C. Fehsenfeld, K.M. Evensen and H.P. Broida, Rev. Sci. Instrum., 36 (1965) 294.
110 H.P. Broida and M.W. Chapman, Anal. Chem., 30 (1958) 2049.
111 R.K. Skogerboe and G.N. Coleman, Anal. Chem., 48 (1976) 611A.
112 F.E. Lichte and R.K. Skogerboe, Anal. Chem., 45 (1973) 399.
113 B. McCarrol, Rev. Sci. Instrum., 41 (1970) 279.

281

114 C.A. Bache and D.J. Lisk, Anal. Chem., 40 (1968) 2224.
115 J.P.S. Haarsma, G.J. de Jong and J. Agterdenbos, Spectrochim. Acta, Part B, 29 (1974) 1.
116 C.I.M. Beenakker, Spectrochim. Acta Part B, 31 (1976) 483.
117 H. Kawaguchi and B.L. Vallee, Anal. Chem., 47 (1975) 1029.
118 R.M. Dagnall, S.J. Pratt, T.S. West and D.R. Deans, Talanta, 16 (1969) 797.
119 C.I.M. Beenakker, Spectrochim. Acta Part B, 32 (1977) 173.
120 D. Alger, D.J. Johnson and G.F. Kirkbright, 20th Coll. Spectrosc. Int. and 7th Int. Conf. At. Spectrosc., Prag, September 1977.
121 F.L. Fricke, O. Rose and J.A. Caruso, Anal. Chem., 47 (1975) 2018.
122 O. Rose, D.W. Mincey, A.M. Yacynych, W.R. Heineman and J.A. Caruso, Analyst, 101 (1976) 753.
123 F.L. Fricke, O. Rose and J.A. Caruso, Talanta, 23 (1976) 317.
124 K.M. Aldous, R.M. Dagnall, B.L. Sharp and T.S. West, Anal. Chim. Acta, 54 (1971) 233.
125 H. Kawaguchi and D.S. Auld, Clin. Chem., 21 (1975) 591.
126 H. Kawaguchi, M. Hasegawa and A. Mizuike, J. Jpn. Spectrosc. Soc., 21 (1972) 36.
127 L.R. Layman and G.M. Hieftje, Anal. Chem., 47 (1975) 194.
128 F.E. Lichte and R.K. Skogerboe, Anal. Chem., 44 (1972) 1321.
129 F.A. Serravallo and T.H. Risby, Anal. Chem., 48 (1976) 673.
130 H.E. Taylor, J.H. Gibson and R.K. Skogerboe, Anal. Chem., 42 (1970) 876.
131 H. Kawaguchi, M. Hasegawa and A. Mizuike, Spectrochim. Acta Part B, 27 (1972) 205.
132 K.F. Fallgatter, V. Svoboda and J.D. Winefordner, Appl. Spectrosc., 25 (1971) 347.
133 D.N. Hingle, G.F. Kirkbright and R.M. Bailey, Talanta, 16 (1969) 1223.
134 F.W. Lampe, T.H. Risby and F.A. Serravallo, Anal. Chem., 49 (1977) 560.
135 H. Kawaguchi, I. Atsuya and B.L. Vallee, Anal. Chem., 49 (1977) 266.
136 K.W. Busch and T.J. Vickers, Spectrochim. Acta Part B, 28 (1973) 85.
137 P. Brassem, F.J.M.J. Maessen and L. de Galan, Spectrochim. Acta Part B, 31 (1976) 537.
138 P. Brassem and F.J.M.J. Maessen, Spectrochim. Acta Part B, 29 (1974) 203.
139 E.V. Belousov, A.S. Bryukhovetskii and A.V. Prokopov, Zh. Tekh. Fiz., 44 (1974) 91.
140 R.M. Dagnall and B.L. Sharp, Proc. Soc. Anal. Chem., 10 (1973) 89.
141 J. Moore, T.S. West and R.M. Dagnall, Proc. Soc. Anal. Chem., 10 (1973) 197.
142 S. Murayama, Spectrochim. Acta Part B, 25 (1970) 191.
143 R.K. Skogerboe and G.N. Coleman, Appl. Spectrosc., 30 (1976) 504.
144 G. Kaiser, D. Götz, G. Tölg, G. Knapp, B. Maichin and H. Spitzy, Z. Anal. Chem., 291 (1978) 278.
145 W.R. Hatch and W.L. Ott, Anal. Chem., 40 (1968) 2085.
146 G. Volland, P. Tschöpel and G. Tölg, Anal. Chim. Acta, 90 (1977) 15.
147 G. Volland, P. Tschöpel and G. Tölg, Z. Anal. Chem., in press.
148 F.E. Lichte and R.K. Skogerboe, Anal. Chem., 44 (1972) 1480.

149 T. Sakamoto, H. Kawaguchi and A. Mizuike, Bunseki Kagaku, 24 (1975) 457.
150 R.J. Watling, Anal. Chim. Acta, 75 (1975) 281.
151 H.E. Taylor, J.H. Gibson and R.K. Skogerboe, Anal. Chem., 42 (1970) 1569.
152 W.D. Hagenah, K. Laqua and F. Leis, XVII. Coll. Spectrosc. Int., Florenz, 1973. Proc. CSI Acta, Vol. II, p. 491.
153 H. Kawaguchi, M. Okada and A. Mizuike, Bunseki Kagaku, 25 (1976) 344.
154 T. Sakamoto, H. Kawaguchi and A. Mizuike, Bunko Kenkyu, 25 (1976) 35.
155 H.U. Eckert, ICP Inf. Newsl., 2 (1977) 327.
156 R. Avni and J.D. Winefordner, Spectrochim. Acta Part B, 30 (1975) 281.
157 G.E. Bently and M.L. Parsons, Anal. Chem., 49 (1977) 551.
158 S.L. Castleden and G.F. Kirkbright, Proc. Anal. Div. Chem. Soc., 13 (1976) 331.
159 W.B. Barnett, J.W. Vollmer and S.M. De Nuzzo, At. Absorp. Newsl., 15 (1976) 33.
160 A.J. Bodretsova, B.V. L'vov and B.I. Mosichev, Zh. Prikl. Spectrosc., 4 (1966) 207.
161 D.O. Cooke, R.M. Dagnall and T.S. West, Anal. Chim. Acta, 54 (1971) 381.
162 R.M. Dagnall, R. Pribil and T.S. West, Analyst, 93 (1968) 281.
163 R.M. Dagnall, K.C. Thompson and T.S. West, Talanta, 14 (1967) 551.
164 R.M. Dagnall and T.S. West, Appl. Opt., 7 (1968) 1287.
165 R.M. Dagnall, M.R.G. Taylor and T.S. West, Spectrosc. Lett., 1 (1968) 397.
166 B.J. Jansen, T.J. Hollander and L.P. Franken, Spectrochim. Acta Part B, 29 (1974) 37.
167 J.M. Mansfield, M.P. Bratzel, H.O. Norgordon, D.O. Knapp, K.E. Zacha and J.D. Winefordner, Spectrochim. Acta Part B, 23 (1968) 389.
168 K.C. Thompson and P.C. Wildy, Analyst, 95 (1970) 776.
169 P.C. Wildy and K.C. Thompson, Analyst, 95 (1970) 562.
170 C. Woodward, At. Absorp. Newsl., 8 (1969) 121.
171 J. Agterdenbos and P.F.E. van Montfort, 20th Coll. Spectrosc. Int. and 7th Int. Conf. At. Spectrosc., Prag, September 1977.
172 G.B. Cook, J.A. Goleb and V. Middleboe, Nature (London), 216 (1967) 475.
173 M.M. Ferraris and G. Proksch, Anal. Chim. Acta, 59 (1972) 177.
174 J.A. Goleb and V. Middleboe, Anal. Chim. Acta, 43 (1968) 229.
175 J.P. Leicknam, V. Middleboe and G. Proksch, Anal. Chim. Acta, 40 (1968) 487.
176 H. Perschke, E.A. Keroe, G. Proksch and A. Muehl, Anal. Chim. Acta, 53 (1971) 459.
177 C.C. Garber and J.W. Taylor, Anal. Chem., 48 (1976) 2070.
178 P.F.E. van Montfort and J. Agterdenbos, Talanta, 21 (1974) 660.
179 H.U. Eckert, 20th Coll. Spectrosc. Int. and 7th Int. Conf. At. Spectrosc., Prag, September 1977.
180 J. Gatzke, Spectrochim. Acta Part B, 30 (1975) 235.
181 V.A. Gruzdev, High Temp. (USSR), 6 (1968) 921.
182 R. Ishida, Rep. Goot. Chem. Ind. Res. Inst. Tokyo, 51 (1956) 342.
183 R.V. Mitin and K.K. Pryadkin, Sov. Phys. Tech. Phys., 10 (1966) 933.

184 R.V. Mitin, A.V. Zvyaginbev and K.K. Pryadkin, High Temp. (USSR), 11 (1973) 443.
185 C.D. West, Anal. Chem., 42 (1970) 811.
186 A. Karmen and R.L. Bowman, Ann. N.Y. Acad. Sci., 72 (1959) 714.
187 M.A. Hedayati, Bull. Iran. Pet. Inst., 58 (1975) 5.
188 M.B. Denton and D.L. Windsor, 27th Pittsburgh Conf. Anal. Chem. Appl. Spectrosc., ICP Inf. Newsl., 1/11 (1976) 263.
189 D.J. David, Gas Chromatographic Detectors, Wiley-Interscience, New York, 1974.
190 R.M. Dagnall, D.J. Johnson and T.S. West, Spectrosc. Lett., 6 (1973) 87.
191 R.M. Dagnall and P. Whitehead, Proc. Soc. Anal. Chem., 9 (1972) 201.
192 L.S. Ettre, J. Chromatogr., 112 (1975) 1.
193 P.M. Houpt, Anal. Chim. Acta, 86 (1976) 129.
194 D.F.S. Natusch and T.M. Thorpe, Anal. Chem., 45 (1974) 1184A.
195 W.R. McLean, D.L. Stanton and G.E. Penketh, Analyst, 98 (1973) 432.
196 W.R. McLean, D.L. Stanton and G.E. Penketh, Analyst, 98 (1973) 377.
197 B.J. Lowing, Analusis, 1 (1972) 510.
198 A. Shilman, Environ. Lett., 6 (1974) 149.
199 R.M. Dagnall, T.S. West and P. Whitehead, Anal. Chim. Acta, 60 (1972) 25.
200 F.A. Serravallo and T.H. Risby, J. Chromatogr. Sci., 12 (1974) 585.
201 C.A. Bache and D.J. Lisk, Anal. Chem., 39 (1967) 786.
202 C.A. Bache and D.J. Lisk, Anal. Chem., 38 (1966) 1757.
203 C.A. Bache and D.J. Lisk, Anal. Chem., 37 (1965) 1477.
204 Y. Talmi and V.E. Norvell, Anal. Chim. Acta, 85 (1976) 203.
205 C.I.M. Beenakker and P.W.J.M. Boumans, 20th Coll. Spectrosc. Int. and 7th Int. Conf. At. Spectrosc., Prag, September 1977.
206 H.A. Moye, Anal. Chem., 39 (1967) 1441.
207 F.A. Serravallo and T.H. Risby, Anal. Chem., 47 (1975) 2141.
208 J.P.J. van Dalen, P.A. de Lezenne Coulander and L. de Galan, Anal. Chim. Acta, 94 (1977) 1.
209 J.P.J. van Dalen, P.A. de Lezenne Coulander and L. de Galan, 20th Coll. Spectrosc. Int. and 7th Int. Conf. At. Spectrosc., Prag, September 1977.
210 M.S. Black and R.E. Sievers, Anal. Chem., 48 (1976) 1872.
211 H. Kawaguchi, T. Sakamoto and A. Mizuike, Talanta, 20 (1973) 321.
212 R.M. Dagnall, T.S. West and P. Whitehead, Analyst, 98 (1973) 647.
213 W.E.L. Grossman, J. Eng and Y.C. Tong, Anal. Chim. Acta, 60 (1972) 447.
214 P.M. Houpt and H. Compaan, Analusis, 1 (1972) 27.
215 Y. Talmi, Anal. Chim. Acta, 74 (1975) 107.
216 C.A. Bache and D.J. Lisk, Anal. Chem., 43 (1971) 950.
217 C.A. Bache and D.J. Lisk, Biomed. Appl. Gas Chromatogr., 2 (1968) 165.
218 C.A. Bache and D.J. Lisk, J. Assoc. Off. Anal. Chem., 50 (1967) 1246.
219 C.A. Bache and D.J. Lisk, Anal. Chem., 38 (1966) 783.
220 C.A. Bache and D.J. Lisk, J. Gas Chromatogr., 6 (1968) 301.
221 R.M. Dagnall, S.J. Pratt, T.S. West and D.R. Deans, Talanta, 17 (1970) 1009.
222 R.M. Dagnall, T.S. West and P. Whitehead, Anal. Chem., 44 (1972) 2074.
223 Y. Talmi and D.T. Bostik, Anal. Chem., 47 (1975) 2145.
224 Y. Talmi and V.E. Norvell, Anal. Chem., 47 (1975) 1510.

225 T. Sakamoto, H. Kawaguchi and A. Mizuike, Jpn Anal., 25 (1976) 80.
226 Y. Talmi and A.W. Andren, Anal. Chem., 46 (1974) 2122.
227 T. Sakamoto, M. Okada, H. Kawaguchi and A. Mizuike, Jpn. Anal., 25 (1976) 85.
228 L.E. Boos and J.D. Winefordner, Anal. Chem., 44 (1972) 1020.
229 W. Braun, N.C. Peterson, A.M. Bass and M.G. Karylo, J. Chromatogr., 55 (1971) 237.
230 D.A. Luippold and J.L. Beauchamp, Anal. Chem., 42 (1970) 1374.
231 T. Sakamoto, H. Kawaguchi and A. Mizuike, J. Chromatogr., 121 (1976) 383.
232 L.A. Shapunov, N.G. Vil'dt, S.D. Okhrimets and A.L. Vil'dt, Zavod. Lab., 39 (1973) 959.
233 W. Kessler and F. Gebhardt, Glastech. Ber., 40 (1967) 194.
234 W. Kessler, Glastech. Ber., 44 (1971) 479.
235 F. Gebhardt and H. Horn, Glastech. Ber., 44 (1971) 483.
236 L.G. Bachurina, I.I. Devyatkin, V.M. Perminova and S.A. Savostin, Zavod. Lab., 39 (1973) 225.
237 S. Murayama, J. Appl. Phys., 39 (1968) 5478.
238 S. Murayama, Bunko Kenkyu, 19 (1970) 237.
239 S. Murayama, H. Matsumo and M. Yamamoto, Spectrochim. Acta Part B, 23 (1967/68) 513.
240 H. Goto, K. Hirokawa and M. Suzuki, Z. Anal. Chem., 225 (1967) 130.
241 S. Greenfield, Metron (Rovigo), 3 (1971) 224.
242 K. Kitagawa and T. Takeuchi, Anal. Chim. Acta, 60 (1972) 309.
243 K. Kitagawa and T. Takeuchi, Anal. Chim. Acta, 67 (1973) 453.
244 R. Mavrodineanu and H. Boiteux, Flame Spectroscopy, Wiley, New York, 1965, p. 85.
245 J.E. Meinhard, ICP Inf. Newsl., 2 (1976) 163.
246 R.N. Kniseley, H. Amenson, C.C. Buttler and V.A. Fassel, Appl. Spectrosc., 28 (1974) 285.
247 J. Dahmen and H. Hölzel, 10. Spektrometertagung-Kurzreferate, Den Haag, 1974, p. 42.
248 E. Jackwerth, Spectrochim. Acta Part B, 30 (1975) 169.
249 K.A. Egorova, Zh. Prikl. Spektrosk., 6 (1976) 22.
250 K.A. Egorova and A.J. Perevertun, Izv. Sib. Otd. Akad. Nauk SSSR Ser. Khim. Nauk, (1967) 132.
251 S. Lanz, W. Lochte-Holtgreven and G. Traving, Z. Phys., 176 (1963) 1.
252 P. Brassem and F.J.M.J. Maessen, Spectrochim. Acta Part B, 30 (1975) 547.
253 I. Atsuya and K. Akatsuka, Anal. Chim. Acta, 81 (1976) 61.
254 I. Atsuya, Anal. Chim. Acta, 74 (1975) 1.
255 H. Horn, F. Gebhardt and W. Kessler, XVI. Coll. Spectrosc. Int., Vol. II, p. 291.
256 K. Govindaraju, G. Mevelle and C. Chonard, Anal. Chem., 48 (1976) 1325.
257 G. Kölblin, U. Stix, E. Grallath, P. Tschöpel and G. Tölg, Z. Anal. Chem., in press.
258 W.J. Maeck, M.E. Kussy, B.E. Ginther, G.V. Wheler and J.E. Rein, Anal. Chem., 35 (1963) 62.
259 R. Nakashima and S. Sasaki, Anal. Chim. Acta, 85 (1976) 75.

260 R. Nakashima, S. Sasaki and S. Shibata, Anal. Chim. Acta, 70 (1974) 265.
261 R. Nakashima, S. Sasaki and S. Shibata, Anal. Chim. Acta, 77 (1975) 65.
262 M. Sermin, Analusis, 2 (1973) 186.
263 H. Kaiser, Spectrochim. Acta, 3 (1947) 40.
264 L.G. Bachurina, I.I. Devyatkin, V.M. Perminova and N.I. Tsemko, Zh. Prikl. Spektrosk., 15 (1971) 401.
265 R. Nakashima, Jpn. Anal., 25 (1976) 869.
266 M. Suzuki, Jpn. Anal., 17 (1968) 1529.
267 M. Suzuki, Jpn. Anal., 19 (1970) 204.
268 M. Suzuki, Jpn. Anal., 19 (1970) 207.
269 J.F. Alder and J.M. Mermet, Spectrochim. Acta Part B, 28 (1973) 421.
270 H.U. Eckert, High Temp. Sci., 6 (1974) 99.
271 C. Veillon and M. Margoshes, Spectrochim. Acta Part B, 23 (1968) 503.
272 S. Greenfield, I.Ll. Jones, H.McD. McGeachin and P.B. Smith, Anal. Chim. Acta, 74 (1975) 225.
273 V.A. Fassel and R.N. Kniseley, Anal. Chem., 46 (1974) 1110A.
274 V.A. Fassel and R.N. Kniseley, Anal. Chem., 46 (1974) 1155A.
275 P.W.J.M. Boumans, F.J. de Boer and J.W. de Ruiter, Philips Tech. Rundsch., 33 (1973/74) 51.
276 S. Greenfield, 1st Eur. ICP Symp., Munich, 1976.
277 R.M. Barnes and R.G. Schleicher, Spectrochim. Acta Part B, 30 (1975) 109.
278 D.J. Kalnicky, R.N. Kniseley and V.A. Fassel, Spectrochim. Acta Part B, 30 (1975) 511.
279 A.R. Apsit, Zh. Tekh. Fiz., 40 (1970) 1527.
280 R.M. Barnes and S. Nikdel, Appl. Spectrosc., 29 (1975) 477.
281 R.M. Barnes and S. Nikdel, J. Appl. Phys., 47 (1976) 3929.
282 R.M. Barnes and S. Nikdel, Appl. Spectrosc., 30 (1976) 310.
283 R.M. Barnes and R.G. Schleicher, Anal. Chem., 46 (1974) 1342.
284 R.M. Barnes and R.G. Schleicher, 16th Coll. Spectrosc. Int. 1971, Vol. II, p. 297.
285 H.U. Eckert, J. Appl. Phys., 41 (1970) 1520.
286 H.U. Eckert, D.C. Pridmore-Brown, J. Appl. Phys., 42 (1971) 5051.
287 J. Jarosz, J.M. Mermet and J.C.R. Robin, C.R. Acad. Sci. Paris, 278 (1974) 885.
288 H.G.C. Human and R.H. Scott, Spectrochim. Acta Part B, 31 (1976) 459.
289 D.J. Kalnicky, F.A. Fassel and R.N. Kniseley, Appl. Spectrosc., 31 (1977) 137.
290 G. Kornblum, ICP Inf. Newsl., 2 (1976) 79.
291 G.R. Kornblum and L. de Galan, Spectrochim. Acta Part B, 32 (1977) 71.
292 T.E. Edmonds and G. Horlick, Appl. Spectrosc., 31 (1977) 536.
293 G.R. Kornblum and L. de Galan, Spectrochim. Acta Part B, 29 (1974) 249.
294 J.M. Mermet, C.R. Acad. Sci. Ser. B, 281 (1975) 273.
295 J.M. Mermet, C.R. Acad. Sci. Ser. B, 284 (1977) 319.
296 J.M. Mermet and J.P. Robin, Rev. Int. Hautes Temp. Refract., 10 (1973) 133.
297 J.M. Mermet, Spectrochim. Acta Part B, 30 (1975) 383.
298 D.C. Pridmore-Brown, J. Appl. Phys., 41 (1970) 3621.

299 Yu.P. Raizer, Sov. Phys. Usp., 12 (1970) 777.
300 B. Talayrach, J. Besombes-Vailhe and H. Triche, Analusis, 1 (1972) 135.
301 P.W.J.M. Boumans and F.J. de Boer, 20th Coll. Spectrosc. Int. and 7th Int. Conf. At. Spectrosc., Prag, 1977.
302 M.E. Britske, Yu.S. Sukach, L.N. Filimonov, Zh. Prikl. Spektrosk., 25 (1976) 5.
303 P.W.J.M. Boumans and F.J. de Boer, Spectrochim. Acta Part B, 27 (1972) 391.
304 G.F. Kirkbright, A.F. Ward and T.S. West, Anal. Chim. Acta, 62 (1972) 241.
305 Ed. Valvo Unernehmensbereich Bauelemente der Philips GmbH, Hochfrequenz-Industrie-Generatoren, Boysen und Maasch, Hamburg, 1975.
306 R.G. Schleicher and R.M. Barnes, Anal. Chem., 47 (1975) 724.
307 H. Linn, ICP Inf. Newsl., 2 (1976) 51.
308 M.H. Abdallah, J. Jarosz, J.M. Mermet, C. Trassy and J. Robin, ICP Inf. Newsl., 1 (1975) 103.
309 P.W.J.M. Boumans and F.J. de Boer, Spectrochim. Acta Part B, 32 (1977) 365.
310 P.W.J.M. Boumans and F.J. de Boer, Spectrochim. Acta Part B, 30 (1975) 309.
311 P.W.J.M. Boumans and F.J. de Boer, Spectrochim. Acta Part B, 31 (1976) 355.
312 H.R. Sobel, R.N. Knisely, W.L. Sutherland and V.A. Fassel, Pittsburgh Conf. Anal. Chem. Appl. Spectrosc., Cleveland, 1975.
313 P.B. Smith and S. Greenfield, ICP Inf. Newsl., 1 (1976) 217.
314 R.H. Scott, A. Strasheim and M.L. Kokot, Anal. Chim. Acta, 82 (1976) 67.
315 R.H. Scott and A. Strassheim, Anal. Chim. Acta, 76 (1975) 71.
316 R.H. Scott and M.L. Kokot, ICP Inf. Newsl., 1 (1975) 2, 34.
317 R.H. Scott and M.L. Kokot, Anal. Chim. Acta, 75 (1975) 257.
318 R.H. Scott, V.A. Fassel, R.N. Kniseley and D.E. Nixon, Anal. Chem., 46 (1974) 75.
319 J.M. Mermet, ICP Inf. Newsl., 2 (1976) 70.
320 C.C. Wohlers, Pittsburgh Conf. Anal. Chem. Appl. Spectrosc., Cleveland, 1977.
321 C.C. Wohlers, ICP Inf. Newsl., 2 (1977) 317.
322 K.W. Olson, W.J. Hass and V.A. Fassel, Anal. Chem., 49 (1977) 632.
323 K. Ohls, ICP Inf. Newsl., 2 (1977) 357.
324 J.M. Mermet and J.P. Robin, Anal. Chem., 40 (1968) 1918.
325 C. Veillon and M. Margoshes, Spectrochim. Acta Part B, 23 (1968) 553.
326 C.T. Apel, T.M. Bieniewski, L.E. Cox and D.W. Steinhaus, ICP Inf. Newsl., 3 (1977) 1.
327 K. Ohls, ICP Inf. Newsl., 1 (1976) 278.
328 R.N. Kniseley, ICP Inf. Newsl., 1 (1975) 92.
329 H.C. Hoare and R.A. Mostyn, Anal. Chem., 39 (1967) 1153.
330 R.H. Scott, Analusis, 4 (1976) 323.
331 M.H. Abdallah, J.M. Mermet and C. Trassy, Anal. Chim. Acta, 87 (1976) 329.
332 R.L. Dahlquist, J.W. Knoll and R.E. Hoyt, Pittsburgh Conf. Anal. Chem. Appl. Spectrosc., Cleveland, 1975.

333 D.J. Koop, M.D. Silvester and J.C. van Loon, ICP Inf. Newsl., 2 (1977) 301.
334 A.F. Ward, ICP Inf. Newsl., 2 (1977) 303.
335 R.N. Kniseley, V.A. Fassel, C.C. Butler, Clin. Chem., 19 (1973) 807.
336 D.E. Nixon, V.A. Fassel and R.N. Kniseley, Anal. Chem., 46 (1974) 210.
337 I. Kleinmann and V. Svoboda, Anal. Chem., 41 (1969) 1029.
338 H.G.C. Human, R.H. Scott, A.R. Oakes and C.D. West, Analyst, 101 (1976) 265.
339 A.S. Bazhov, V.K. Zakharov, P.A. Koka and A.F. Malinovskaya, Zavod. Lab., 34 (1968) 245.
340 R.D. Gerasimov and G.S. Eilenkrig, Zh. Prikl. Spektrosk., 19 (1973) 791.
341 T.B. Reed, J. Appl. Phys., 32 (1961) 821.
342 T.B. Reed, J. Appl. Phys., 32 (1961) 2534.
343 T.B. Reed, Int. Sci. Technol., 42 (1962) 42.
344 G. Pforr and O. Aribot, Z. Chem., 10 (1970) 78.
345 R.M. Dagnall, D.J. Smith, T.S. West and S. Greenfield, Anal. Chim. Acta, 54 (1971) 397.
346 B. Bogdain, ICP Inf. Newsl., 2 (1977) 275.
347 J.L. Genna, R.M. Barnes and C.D. Allemand, Anal. Chem., 49 (1977) 1450.
348 Ann. Rep. Anal. Spectrosc., 5 (1976) 8.
349 I.P. Dashkevich and G.S. Eilenkrig, Zavod. Lab., 39 (1973) 415.
350 A. Czernichowski and J. Jurewicz, Pr. Nauk. Inst. Chem. Nieorg. Metal. Pierwiastkow Politech. Wroclaw., 24 (1975) 3.
351 P.W.J.M. Boumans, G.H. van Gool, J.A.J. Jansen, 11 Spektrometertagung, Montreux, Schweiz, 1976.
352 P.W.J.M. Boumans and F.J. de Boer, 20th Coll. Spectrosc. Int. and 7th Int. Conf. At. Spectrosc., Prag, 1977.
353 C. Allemand and G. Benzie, ICP Inf. Newsl., 1 (1976) 273.
354 C. Allemand, ICP Inf. Newsl., 2 (1976) 68.
355 D. Truitt and J.W. Robinson, Anal. Chim. Acta, 49 (1970) 401.
356 C. Allemand, 27th Pittsburgh Conf. Anal. Chem. Appl. Spectrosc., Cleveland, 1975.
357 R.M. Ajhar, P.D. Dalager and A.L. Davison, Pittsburgh Conf. Anal. Chem. Appl. Spectrosc., Cleveland, 1975.
358 P.W.J.M. Boumans, ICP Inf. Newsl., 1 (1975) 68.
359 P.W.J.M. Boumans, Spectrochim. Acta Part B, 31 (1976) 147.
360 P.W.J.M. Boumans, ICP Inf. Newsl., 2 (1976) 101.
361 S. Greenfield, H.McD. McGeachin and P.B. Smith, Anal. Chim. Acta, 84 (1976) 67.
362 S. Greenfield, H.McD. McGeachin and P.B. Smith, ICP Inf. Newsl., 2 (1976) 167.
363 G.F. Kirkbright, A.F. Wart and T.S. West, Anal. Chim. Acta, 64 (1973) 353.
364 J.M. Mermet and J. Robin, Anal. Chim. Acta, 75 (1975) 271.
365 G.F. Larson, V.A. Fassel, R.H. Scott and R.N. Kniseley, Anal. Chem., 47 (1975) 238.
366 V.A. Fassel, C.A. Peterson, F.N. Abercrombie and R.N. Kniseley, Anal. Chem., 48 (1976) 516.

288

367 G.F. Larson and V.A. Fassel, Anal. Chem., 48 (1976) 1161.
368 C.C. Butler, R.N. Kniseley and V.A. Fassel, Anal. Chem., 47 (1975) 825.
269 J.C. Souilliart and J.P. Robin, Analusis, 1 (1972) 427.
370 D. Truitt and J.W. Robinson, Anal. Chim. Acta, 51 (1970) 61.
371 I.A. Maiorov, L.N. Filimonov and B.V. Yarkin, Zh. Prikl. Spektrosk., 22 (1975) 561.
372 I. Kleinmann and V. Svoboda, Radioisotopy, 15 (1974) 7.
373 A.F. Ward, H.R. Sobel and R.L. Crawford, 3rd Ann. FACSS Mtg., Philadelphia, 1976.
374 V.N. Soshnikov, E.S. Trekhov and Yu.M. Khoshev, Zh. Prikl. Mekh. Tekh. Fiz., 1 (1971) 148.
375 D.E. Nixon and V.A. Fassel, 27th Pittsburgh Conf. Anal. Chem. Appl. Spectrosc., Cleveland, 1976.
376 B.T.N. Newland and R.A. Mostyn, ICP Inf. Newsl., 2 (1976) 135.
377 K. Ohls, K.H. Koch and H. Grote, Z. Anal. Chem., 284 (1977) 177.
378 K. Ohls, First Eur. ICP Symp. Munich, June, 1976.
379 B.T.N. Newland and R.A. Mostyn, ICP Inf. Newsl., 1 (1976) 183.
380 R.N. Kniseley, 5th Ann. Symp. Recent Adv. Anal. Chem. Pollutants, Jekyll Island, Georgia, May, 1975.
381 R.N. Kniseley, Rep. At. Energy Comm. US, 1974, JS-T-626, 89 pp.
382 J. Jarosz and J.M. Mermet, J. Quant. Spectrosc. Radiat. Transfer, 17 (1977) 237.
383 J.B.Jones, Commun. Soil Sci. Plant. Anal., 8 (1977) 349.
384 S.D. Gupta, M.D. Silvester and R.B. Cruz, 27th Pittsburgh Conf. Anal. Chem. Appl. Spectrosc., Cleveland, 1976.
385 A. Danielsson, 27th Pittsburgh Conf. Anal. Chem. Appl. Spectrosc., Cleveland, 1976.
386 A. Danielsson and E. Soderman, ICP Inf. Newsl., 2 (1977) 267.
387 S.E. Church, ICP Inf. Newsl., 2 (1977) 314.
388 R.L. Dahlquist and J.W. Knoll, ICP Inf. Newsl., 2 (1977) 313.
389 F. Brech, ICP Inf. Newsl., 1 (1976) 195.
390 F. Breck and R.L. Crawford, Pittsburgh Conf. Anal. Chem. Appl. Spectrosc., Cleveland, 1975.
391 J.A.C. Broekaert, F. Leis and K. Laqua, 20th Coll. Spectrosc. Int. and 7th Int. Conf. At. Spectrosc., Prag, 1977.
392 J.O. Burman, ICP Inf. Newsl., 3 (1977) 33.
393 P.W.J.M. Boumans, G.H. van Gool and J.A.J. Jansen, Analyst, 101 (1976) 585.
394 P.W.J.M. Boumans, ICP Inf. Newsl., 1 (1976) 206.
395 F.N. Abercrombie and M.D. Silvester, Pittsburgh Conf. Anal. Chem. Appl. Spectrosc., Cleveland, 1975.
396 R.M. Ajhar, P.D. Dalanger and A.L. Davison, Am. Lab., 8 (1976) 71.
397 C. Allemand, ICP Inf. Newsl., 2 (1976) 1.
398 F.N. Abercrombie and M.D. Silvester, 27th Pittsburgh Conf. Anal. Chem. Appl. Spectrosc., Cleveland, 1976.
399 F.N. Abercrombie, M.D. Silvester and G.S. Stonte, ICP Inf. Newsl., 2 (1977) 309.

400 D.W. Golightly and J.L. Harris, Appl. Spectrosc., 29 (1975) 233.
401 I. Kleinmann and V. Svoboda, Chem. Listy, 69 (1975) 833.
402 R. Lerner, Spectrochim. Acta, 20 (1964) 1619.
403 D.R. Marriott, Proc. Soc. Anal. Chem., 10 (1973) 90.
404 T. Takeuchi, Ashi Garasu Kogyo Gitutsu Shorei-Kai, Kenkyn Kokoku, 12 (1966) 199.
405 V.S. Engelst, K.U. Urmanbetov, Z.Z. Zeenbaev, Zavod. Lab., 42 (1976) 174.
406 M. Yamamoto, Jpn. J. Appl. Phys., 1 (1962) 235.
407 M. Yamamoto, Jpn. J. Appl. Phys., 2 (1963) 62.
408 M. Yamamoto, Jpn. J. Appl. Phys., 2 (1963) 410.
409 S.E. Valente and W.G. Schrenk, Appl. Spectrosc., 24 (1970) 197.
410 E.H. Sirois, Anal. Chem., 36 (1964) 2389.
411 E.H. Sirois, Anal. Chem., 36 (1964) 2394.
412 R.K. Skogerboe, I.T. Urasa and G.N. Coleman, Appl. Spectrosc., 30 (1976) 500.
413 D.M. Shaw, O. Wickremasinghe and C. Yip, Spectrochim. Acta, 13 (1958) 197.
414 M. Riemann, Z. Anal. Chem., 215 (1966) 407.
415 K.H. Neeb and W. Gebauhr, Z. Anal. Chem., 190 (1962) 92.
416 I. Miyachi and K. Jayaram, Rev. Sci. Instrum., 42 (1971) 1002.
417 M. Marinković, Bull. Boris Kidric Inst. Nucl. Sci., 16 (1965) 65.
418 M. Margoshes and B.F. Scribner, J. Res. Nat. Bur. Stand. Sect. A, 67 (1963) 561.
419 M. Marinković and B. Dimitrijević, Spectrochim. Acta, 23 (1968) 257.
420 H. Maecker, Z. Naturforsch. Teil A, 11 (1956) 457.
421 M. Kubota, Anal. Chim. Acta, 88 (1977) 79.
422 E. Kranz, in R. Ritschl and H. Holds (Eds.), Emissionsspektroskopie, Akademie-Verlag, Berlin, 1964, p. 160.
423 E. Kranz, Chem. Anal. (Warsaw), 14 (1969) 1207.
424 V.V. Korolev, N.T. Shokina and M.E. Shuvalova, Zh. Anal. Khim., 31 (1976) 1891.
425 F.A. Korolev and Ju.K. Kvaraccheli, Opt. Spektrosk., 10 (1961) 398.
426 M. Kliska and M. Marinković, Spectrochim. Acta Part B, 25 (1970) 545.
427 L. Klein, Rev. Sci. Instrum., 41 (1970) 668.
428 R.J. Heemstra, Dev. Appl. Spectrosc., 9 (1971) 199.
429 H. Goto and I. Atsuya, Z. Anal. Chem., 225 (1967) 121.
430 W.G. Elliot, Am. Lab., 3 (1972) 45.
431 K. Doerffel and J. Lichtner, Spectrochim. Acta, 22 (1966) 1245.
432 J.F. Chapman, L.S. Dale and R.N. Whittem, Analyst, 98 (1973) 529.
433 F. Burhorn, H. Maecker and T. Peters, Z. Phys., 131 (1951) 28.
434 V. Svoboda and I. Kleinmann, Radiosiotopy, 15 (1974) 827.
435 P. Merchant and C. Veillon, Anal. Chim. Acta, 70 (1974) 17.
436 L.E. Owen, AFML-TR-67-400, Air Force Materials Laboratory, Wright-Patterson Air Force Base, Ohio, 1967.
437 E. Kranz, Wiss. Z. Tech. Hochsch. Ilmenau, 16 (1970) 181.
438 E. Kranz, Spectrochim. Acta Part B, 27 (1972) 327.

439 Z. Zheenbaev, F.G. Karich, R.J. Konakov, K. Urmanbetov and V.S. Engelst, Zavod. Lab., 35 (1969) 1343.
440 H. Raab and W. Bogerhausen. Spectrochim. Acta Part B, 25 (1970) 183.
441 K. Acinger, D. Miler, G. Pichler, A. Mesjaski-Toneje and V. Vujnović, Fizika (Zagreb) Suppl., 2 (1970) 85.
442 H. Raab, J. Phys. E., 5 (1972) 779.
443 I.G. Yudelevich and A.S. Cherevko, Spectrochim. Acta Part B, 31 (1976) 93.
444 R. Zottmann and H. Krempl, Spectrochim. Acta Part B, 26 (1971) 451.
445 G. Holdt and E. Hoffmann, Z. Anal. Chem., 215 (1966) 114.
446 M. Marinković and T.J. Vickers, Appl. Spectrosc., 25 (1971) 319.
447 M. Marinković and M. Kliska, Proc. 15th Colloq. Spectrosc. Int., Madrid, Vol. 4, Hilger, London, 1969, p. 407.
448 B. Zmbova and M. Marinković, Talanta, 20 (1973) 647.
449 L.E. Owen, Appl. Spectrosc., 15 (1961) 150.
450 E.N. Savinova and A.V. Karyakin, Z. Anal. Chem., 25 (1970) 1379.
451 E.N. Savinova, A.V. Karyakin and T.P. Andreeva, Zh. Anal. Khim., 27 (1972) 777.
452 P.M. McElfresh and M.L. Parsons, Anal. Chem., 46 (1974) 1021.
453 I.G. Yudelevich, A.S. Cherevko and E.S. Bogdanova, Izv. Sib. Otd. Akad. Nauk SSSR, Ser. Khim. Nauk, 4 (1972) 83.
454 M.S.W. Weeb and P.C. Wildy, Nature (London), 198 (1963) 1218.
455 W.E. Rippetoe and T.J. Vickers, Anal. Chem., 47 (1975) 2082.
456 J. Szivek, C. Jones, E.J. Paulson and L.S. Valberg, Appl. Spectrosc., 22 (1968) 195.
457 I. Atsuya, J. Chem. Soc. Jpn., 86 (1965) 1145.
458 I. Atsuya and H. Goto, Spectrochim. Acta Part B, 26 (1971) 359.
459 I. Atsuya and H. Goto, Anal. Chim. Acta, 65 (1973) 303.
460 H. Goto and I. Atsuya, Z. Anal. Chem., 240 (1968) 102.
461 M. Yamamoto, J. Spectrosc. Soc. Jpn., 2 (1963) 84.
462 T. Takeuchi and Y. Katsuno, Jpn. Analyst., 18 (1969) 62.
463 E. Kranz, Proc. 12th Colloq. Spectrosc. Int., Exeter, Hilger, London, 1965, p. 574.
464 E. Kranz, Proc. 14th Colloq. Spectrosc. Int., Debrecen, Vol. 2, Hilger, London, 1968, p. 697.
465 E. Kranz, Proc. 15th Colloq. Spectrosc. Int., Madrid, Vol. 4, Hilger, London, 1969, p. 95.
466 M. Oku and K. Hirokawa, Bull. Chem. Soc. Jpn., 44 (1971) 1010.
467 A.G. Collins and C.A. Pearson, Anal. Chem., 36 (1964) 787.
468 A.G. Collins, Appl. Spectrosc., 21 (1967) 16.
469 D.A. Murdick and E.H. Piepmeier, Anal. Chem., 46 (1974) 678.
470 W.E. Rippetoe, E.R. Johnson and T.J. Vickers, Anal. Chem., 47 (1975) 436.
471 M. Marinković and T.J. Vickers, Appl. Spectrosc., 25 (1971) 319.
472 F.G. Karikh, Primen. Plazmatrona Spektrosk., Mater. Vses. Simp., (1970) 158.
473 R. Ishida and M. Kubota, Bunko Kenkyu, 21 (1972) 255.
474 A.S. Cherevko and V.J. Simonova, Zh. Prikl. Spektrosk., 19 (1973) 348.

291

475 K. Hirokava and H. Goto, Bull. Chem. Soc. Jpn., 42 (1969) 693.
476 S.S. Karacki and F.L. Corcoran, Appl. Spectrosc., 27 (1973) 41.
477 I.G. Yudelevich, A.S. Cherevko and N.G. Skobelkina, Zh. Anal. Khim., 27 (1972) 2119.
478 I.G. Yudelevich and A.S. Cherevko, Zh. Prikl. Spektrosk., 17 (1972) 702.
479 L. Vecsernyes, 16th Colloq. Spectrosc. Int., Vol. 2, Hilger, London, 1971, p. 299.
480 A.K. Rusanov and N.T. Batova, J. Anal. Chem. USSR, 20 (1965) 387.
481 W. Bögershausen and O. Consée, Spectrochim. Acta Part B, 25 (1970) 289.
482 H. Isagawa and E. Niki, Bunko Kenkyu, 19 (1970) 368.
483 H. Schirrmeister, Spectrochim. Acta Part B, 23 (1968) 709.
484 H. Kawaguchi, T. Saga and A. Mizuike, Bunko Kenkyu, 24 (1975) 99.
485 F.L.Jr. Corcoran, P.N. Keliher and C.C. Wohlers, Am. Lab., 4 (1972) 51.
486 L.S. Zakharov, Zh. Prikl. Spektrosk., 18 (1973) 11.
487 I.G. Yudelevich and A.S. Cherevko, Izv. Sib. Otd. Akad. Nauk SSSR, Ser. Khim. Nauk, 4 (1971) 91.
488 E.E. Vainshtein, V.V. Korolev and E.N. Savinova, Zh. Anal. Khim., 16 (1962) 532.
489 A. Miyazaki, A. Kimura and Y. Umezaki, Anal. Chim. Acta, 90 (1977) 119.
490 V.S. Krivchenkova, L.M. Fedorova and A.D. Khakhaev, Zh. Prikl. Spektrosk., 12 (1970) 970.
491 R.J. Heemstra and N.G. Foster, Anal. Chem., 38 (1966) 492.
492 W.A. Loseke and E.L. Grove, Appl. Spectrosc., 26 (1972) 527.
493 I. Kleinmann, Radioisotopy, 12 (1971) 1.
494 I. Kleinmann, Spectrochim. Acta Part B, 27 (1972) 93.
495 I. Kleinmann and J. Cajko, Spectrochim. Acta Part B, 25 (1970) 657.
496 P.D. Johnston and J. Lawton, Nature (London), 230 (1971) 320.
497 D.J. Runser and C.W. Frank, Appl. Spectrosc., 28 (1974) 175.
498 D.J. Runser and C.W. Frank, Anal. Chem., 48 (1976) 514.
499 G. Pforr and V. Kapicha, Collect. Czech. Chem. Commun., 31 (1966) 4710.
500 H. Dunken and G. Pforr, Z. Chem., 6 (1966) 278.
501 O. Scholz, Umsch. Wiss. Tech., 59 (1959) 716.
502 C. Roddy and B. Green, Electron. World, 65 (1961) 29.
503 N.I. Gonchar, A.V. Zvyagintsev, R.V. Mitin and K.K. Pryadkin, Zh. Tekh. Fiz., 45 (1975) 657.
504 J. van Calker and W. Tappe, Arch. Eisenhuettenwes., 34 (1963) 679.
505 G.D. Cristescu and R. Grigorovici, Rev. Phys., 1 (1956) 103.
506 G.D. Cristescu and R. Grigorovici, Rev. Phys., 4 (1959) 153.
507 G.B. Dreher and C.W. Frank, Appl. Spectrosc., 28 (1974) 191.
508 C.L. Chakrabarti, R.J. Magee and C.L. Wilson, Talanta, 9 (1962) 43.
509 T. Given, R.J. Magee and C.L. Wilson, Talanta, 3 (1959) 191.
510 O. Botschkova, S. Frisch and E. Schreider, Spectrochim. Acta, 13 (1958) 50.
511 H. Triche, A. Saadate, B. Talayrach and J. Besombes-Vailhe, Analusis, 1 (1972) 413.
512 H. Triche, A. Saadate and J. Besombes-Vailhe, Methodes Phys. Anal., 8 (1972) 26.

513 A.A. Pupyshev and V.N. Muzgin, Zh. Anal. Khim., 28 (1973) 890.
514 N.G. Snopov, Zh. Prikl. Spektrosk., 18 (1973) 371.
515 M.A. Kabanova and E.N. Sautina, Zh. Prikl. Spektrosk., 25 (1976) 21.
516 H. Triché, M. Tabassian and J. Besombes-Vailhe, Methodes Phys. Anal., 7 (1971) 68.
517 A.A. Muchkaev, V.M. Nemets, A.A. Petrov and G.V. Skvortsova, Zh. Prikl. Spektrosk., 19 (1973) 979.
518 M.E. Britske, V.D. Ignatko and Yu.S. Sukach, Teplofiz. Vys. Temp., 10 (1972) 265.
519 V.I. Aref'ef, S.D. Grishin, L.A. Kuz'min, L.V.Leshkov and V.G. Mikhalev, High Temp. (USSR), 11 (1973) 222.
520 J.M. Mermet and C. Trassy, Appl. Spectrosc., Part B, 31 (1977) 237.
521 M.H. Abdallah, R. Diemiaszonek, J.M. Mermet, J. Robin, C. Trassy and J. Jarosz, Anal. Chim. Acta, 84 (1976) 271.
522 M. Nifuku, H. Isagawa and E. Niki, Jpn. Analyst, 18 (1969) 1364.
523 K.E. Friend and A.J. Diefenderfer, Anal. Chem., 38 (1969) 1763.
524 G.H. Morrison and Y. Talmi, Anal. Chem., 42 (1970) 809.
525 J. Robin, Methodes Phys. Anal., 3 (1971) 3.
526 C. Veillon and M. Margoshes, 13th Colloq. Spectrosc. Int., Ottawa, Hilger, London, 1967, p. 281.
527 M. Malinek and H. Massmann, Can. Spectrosc., 15 (1970) 77.
528 A. Montaser and V.A. Fassel, Anal. Chem., 48 (1976) 1490.
529 A. Montaser, V.A. Fassel and S.A Goldstein, Pittsburgh Conf. Anal. Chem. Appl. Spectrosc., Cleveland, 1975.
530 R.H. Wendt and V.A. Fassel, Anal. Chem., 38 (1966) 337.
531 V.A. Fassel, R.N. Kniseley and C.C. Butler, in O.B. Michelsen (Ed.), Analysis and Application of Rare Materials, Nato Advanced Study Inst. Universitetsforlaget, Oslo, Norway, 1973.
532 ICP Inf. Newsl., 1 (1976) 267.
533 H. Brandenburger, Chimia, 22 (1968) 449.
534 C. Bordonali and M.A. Biancifiori, U.S. Pat. 3,684,884/1972.
535 C. Bordonali and M.A. Biancifiori, Metall. Ital., 59 (1967) 631.
536 C. Bordonali and M.A. Biancifiori, Corsi. Semin. Chim., 9 (1968) 153.
537 C. Bordonali, M.A. Biancifiori and A. Donato, Metall. Ital., 61 (1969) 360.
538 C. Bordonali, M.A. Biancifiori, A. Donato and M. Morello, Comit. Naz. Energia Nucl., 1969, RT/CHI69, p. 18.
539 C. Bordonali and M.A. Biancifiori, 14th Colloq. Spectrosc. Int., Debrecen, Hilger, London, 1968.
540 Y. Talmi and G.H. Morrison, Anal. Chem., 44 (1972) 1455.
541 Y. Talmi and G.H. Morrison, Anal. Chem., 44 (1972) 1467.
542 L.G. Bachurina, V.M. Perminova and S.A. Savostin, Zav. Lab., 40 (1974) 1348.
543 R.D. Irons, E.A. Schenk and R.D. Giaugue, Clin. Chem., 22 (1976) 2018.
544 G.F. Kirkbright, Proc. Anal. Div. Chem. Soc., 12 (1975) 8.
545 G.F. Kirkbright and A.F. Ward, Talanta, 21 (1974) 1145.

Index

Auger electron spectroscopy, absolute detection limit of, 129
—, analysis of nickel—chromium steels by, 128
—, application of, 152—158
—, —, to fracture surfaces, 155—156
—, application of electrostatic analysers in, 104, 107—109
—, application of retarding field analysers in, 104—107
—, application of sputtering in, 112
—, argon ion bombardment in, 110
—, back-scattering factor variations in, 125
—, bake-out conditions for, 115
—, "chemical shifts" in, 122
—, comparison of techniques related to, 161—165
—, cylindrical mirror analysers for, 90, 95, 107—108
—, depth profiling by, 134—148
—, detection of surface contamination by, 153—155
—, determination of overlayer thickness by, 133
—, effect of charging of insulators in, 149
—, effect of surface morphology in, 148—149
—, electron energy analysers for, 94
—, electron gun positioning in, 115
—, electron gun voltage, 116
—, electron-induced artifacts in, 149—150
—, error sources in, 148
—, examination of thin layers by, 129—134
—, fault finding in, 119
—, high energy resolution, 160
—, high lateral resolution, 158
—, historical development of, 89—90
—, instrumentation for, 103—109
—, interface analysis by, 152
—, investigations of chemical reactions by, 124
—, ion beam-induced artifacts in, 151
—, ion pumps for, 113—115
—, limits of detection of, 128
—, loss peaks in, 119
—, operation of spectrometer in, 113—115

—, optimization of elastic peak performance in, 119
—, overmodulation in, 118
—, preparation of standards in, 110—113
—, problems of multielement analysis by, 121—122
—, qualitative analysis by, 119—124
—, quantitative analysis by, 124—129
—, recent advances in data processing in, 160
—, relative elemental sensitivity factors in, 131
—, sensitivity factor in, 126
—, specimen preparation for, 110
—, sputtering profiles measured by, 146
—, total excitation current in, 124
—, ultrahigh vacuum system for, 113—115
Auger intensity ratios, 131
Auger lines, widths of, 98
Auger peak-to-peak heights, 112—113
—, as function of modulation voltage, 116
Auger sensitivity factor, 126—127
Auger signal, magnetic field influence on, 118
Auger spectra electron-induced, 104
—, influence of chemical bonding on, 122
—, ion-induced, 104
—, of electrochemically produced Nb_2O_5, 132
—, overlapping peaks in, 121
—, qualitative analysis by reference to, 120—121
—, recognition of light elements from, 121
—, time dependent shift in, 149
Auger spectrum, of aluminium oxide, 123
—, of bulk tin, 120
—, of carbon, from chromium carbide, 123
—, —, from CO_2 layer, 122
—, of gold, 121
—, of metallic aluminium, 123
—, of niobium, 132
—, of pure copper surface, 111, 117
—, of segregated tin, on copper, 133
—, of silicon, in silicon nitride, 123
—, of sulphur, on gold surface, 121

298

Electrons, detection efficiency of, 54
Electron analyser, calibration of, 51—53
—, collection efficiency of, 60
—, scanning of, 50
—, "zero energy", 55
Electron bombardment, excitation of Auger spectra by, 104
Electron correlation effects, 62, 64
Electron energy analysis, for AES, 104—109
Electron gun, alignment of, for AES, 115
—, voltage of, for AES, 115—116
Electron-induced artifacts, in AES, 149—150
Electron monochromator, 45
Electron multipliers, 53
Electron—optical quality, 104—105
Electron reflection coefficient, 49
Electron scattering, production of equivalent photoelectron spectrum using, 30
Electron spectroscopy, position-sensitive detectors in, 55
Electron synchrotron (see also Synchrotron radiation), 27
Electrostatic analyser, for photoelectron spectroscopy, 40—41
Electrostatic analysers, in AES, 104, 107—109
Emission spectroscopy, excitation sources for, 177
Energy analyser, achievement of field configuration for, 49
—, selection of, 39
—, types of, 40—43
Equilibrium surface segregation, study by AES, 157
Erosion rate, effect of, in depth profiling by AES, 136
ESCA, comparison with AES, 161—162, 165
Escape depth, 94, 95, 96
—, effective mean, 124
—, effect of, on Auger yields, 130
Excitation current, in AES, 124

Fluorides, effects on quartz capillaries, in microwave-induced plasmas, 198

Fluorine, determination in rocks, by plasma jet excitation, 269
Fluorine residues, detection of, by AES of contaminated surfaces, 153
Fraction number, 181
Fracture surfaces, analysis of, by AES, 155—156
Franck—Condon factors, 5, 74—76
Franck—Condon principle, 5—6
Free-running oscillators, 239

Gallium, determination of, by microwave-induced plasma excitation, 191
Gas chromatography, application of microwave-induced plasma excitation to, 210—216
Geological samples, analysis of, using plasma excitation, 256
—, determination of metals in, using d.c. plasma sources, 269
Glass, Auger analysis of, 149
Glow discharge mass spectroscopy, comparison with AES, 163, 164
Glow discharge optical spectroscopy, comparison with AES, 163
Gold, determination of impurities in, by current-carrying d.c. plasmas, 261, 268
Gold—nickel, study of interdiffusion in, by AES, 155
Grain boundary segregation, study of, by AES, 155
Gratings, toroidal, 34
Gryzinski calculation, 101—102

Hafnium, interference of zirconium in determination of, with plasma excitation, 251
Hartree—Fock calculations, for nitrogen, 68
Helium, examination of autoionizing states in, 27
Helium resonance radiation, determination of molecular structure by, 67
—, usable intensity from, 28
Hemispherical analyser, 42, 45—46
High energy resolution AES, 160

High-frequency generators, for inductively coupled plasmas, 237—241
—, precautions required for safe use of, 180—181
High-frequency induced plasmas, 269—273
High-frequency oscillator circuits, 238
High-frequency torch discharge, 271
High lateral resolution AES, 159
Hydrogen, application of photoion kinetic energy spectroscopy to, 80
—, dissociative photoionization of, 78
—, photoelectron band shapes in, 10
Hydrogen bromide, photoelectron band shapes in, 10

Indium, determination of, by microwave-induced plasma excitation, 191
Inductively coupled HF plasma, 179, 182
Inductively coupled microwave plasma, 182, 185
Inductively coupled plasmas, 229—258
—, analytical applications of, 251—258
—, analytical limits of detection with, 251—255
—, background correction for, 247
—, commerical instruments utilizing, 245
—, design characteristics of, 240—241
—, dimensions of, 234
—, establishment of best zone for analysis, 247
—, ignitions of, 246
—, interferences when used as excitation source, 248—250
—, introduction of solid samples into, 244
—, principle of, 229—235
—, residence time of atoms in, 233
—, sample introduction into, 241—244
—, specification of various types of, 232—233
—, tail flame of, 232
—, temperature distribution in, 234
Inductively coupled plasma burner, 235, 236
Inductively coupled plasma generator, 235, 237—241
Inorganic molecules, application of photoelectron spectroscopy to, 67

Inter-element effects, with capacitively coupled microwave plasmas, 225
Inter-element ratios, determination of, by microwave-induced plasma excitation, 216
Interface analysis, by AES, 152
Intergranular fracture studies, by AES, 155
Intershell coupling, 65
Iodine, determination of, by microwave-induced plasma excitation, 191
Ion-beam induced artifacts, in AES, 151
Ion-beam spectrochemical analysis, comparison with AES, 163
Ionization interference, during excitation with inductively coupled plasmas, 249—250
Ionization loss spectroscopy, comparison with AES, 161, 162
Ionization potential, 4, 7
Ion scattering spectroscopy, comparison with AES, 163, 165
Iron, determination in lubricating oil, by plasma excitation, 254, 258
—, determination in orchard leaves, by plasma excitation, 257
—, determination in rocks, using plasma jet excitation, 269
—, determination in stainless steel, by AES, 128
—, determination of, by microwave-induced plasma excitation, 191
—, electrolytic separation of, 201
Iron—nickel, study of interdiffusion in, by AES, 155

Jahn—Teller distortions, 11
Jahn—Teller effects, in photoelectron spectrum of benzene, 69

Klystron, 185
Koopmans' theorem, 68

Lead, determination in lubricating oil, by plasma excitation, 254, 258
—, determination of, by microwave-induced plasma excitation, 191

302